PROSPECTS IN
MATHEMATICS

PROSPECTS IN MATHEMATICS

Invited Talks on the Occasion
of the 250th Anniversary of
Princeton University

March 17–21, 1996
Princeton University

HUGO ROSSI, EDITOR

AMERICAN MATHEMATICAL SOCIETY

1991 *Mathematics Subject Classification*. Primary 00B20.

The photographs of the Fine Hall Faculty and Solomon Lefschetz in John Milnor's article, "Growing up in the Old Fine Hall," pp. 2 and 3, and all of the photographs in the Participants of the Conference section are courtesy of Princeton University.

The photograph of Salomon Bochner in John Milnor's article, "Growing up in the Old Fine Hall," page 4 is courtesy of Princeton University Press.

The photograph on the front and back cover is of Old Fine Hall at Princeton University. It is courtesy of Robert P. Matthews, Communications Department, Princeton University.

Library of Congress Cataloging-in-Publication Data

Prospects in mathematics / Hugo Rossi, editor.

    p. cm.

    Papers from a conference held at Princeton University, Mar. 17–21, 1996.

    Includes bibliographical references.

    ISBN 0-8218-0975-X

    1, Mathematics—Congresses.   I. Rossi, Hugo.

QA1.P792   1998

510–dc21

98-36451

CIP

Contents

Foreword

On March 17–21, 1996, the Mathematics Department of Princeton University held a conference entitled "Prospects in Mathematics" as part of the University's 250th anniversary celebration. These were the invited speakers and the titles of their lectures:

Gerd Faltings, Max-Planck Institüt für Mathematik, "What Do We Know About Diophantine Equations?"

Jürg Fröhlich, Eidgenössische Technische Hochschule Zürich (ETH), "The Electron Is Inexhaustible"

Michael Gromov, Institut des Hautes Études Scientifiques and the University of Maryland, "Qualitative Homotopy Theory"

Richard Hamilton, University of California, San Diego, "Nonlinear Parabolic Partial Differential Equations"

Ehud Hrushovski, Hebrew University of Jerusalem, "Model Theory and Geometry"

Henryk Iwaniec, Rutgers University, "Harmonic Analysis in Number Theory"

Robert Langlands, Institute for Advanced Study, Princeton, "Where Stands Functoriality Today?"

Dusa McDuff, State University of New York at Stony Brook, "Symplectic Topology and Capacities"

John Milnor, State University of New York at Stony Brook, "Growing Up in Old Fine Hall"

Jürgen Moser, ETH, Zürich, "Minimal Foliations in Geometry and Dynamics"

Yum-Tong Siu, Harvard University, "Recent Developments in Several Complex Variables"

Donald C. Spencer, Professor Emeritus, Princeton University, "Old Memories and an Old Problem"

Michael Struwe, ETH, Zürich, "Nonlinear Evolution Problems in Geometry and Physics"

Clifford Taubes, Harvard University, "Geometric Aspects of the Seiberg–Witten Invariants"

Ed Witten, Institute for Advanced Study, Princeton, "Small Instantons in String Theory"

Thomas Wolff, California Institute of Technology, "Combinatorial Questions in L^p Harmonic Analysis"

The invited speakers were asked to present their perspective and views on future developments on their chosen topics. In selecting the list the Organizing Committee did not intend to cover all of the basic parts of mathematics, but rather to choose subjects more or less related to the interests of people in the Department.

As a result of this strategy, we had several dominant themes: geometry (M. Gromov, D. McDuff, C. Taubes, J. Moser), nonlinear partial differential equations (R. Hamilton, M. Struwe), number theory (R. Langlands, H. Iwaniec, G. Faltings), and mathematical physics (J. Fröhlich, E. Witten)and three talks representative of specific significant developments during the past half century (E. Hrushovski, Y.-T. Siu, T. Wolff). The talks by D. Spencer and J. Milnor consisted in large part of reminiscences of their years in Princeton.

We were very fortunate that H. Rossi took on the great task of preparing texts based on videotapes and editing the texts which were sent by the speakers. The Organizing Committee expresses its deep gratitude to H. Rossi for his great help. We also thank the Committee Secretary K. Khanin, the graduate students, the staff and all members of the department who assisted us in preparing and organizing the conference.

The Organizing Committee:

William Browder
Steven Klainerman
Peter Sarnak
Yakov Sinai (Chair)
Elias Stein

Welcoming Remarks

J. J. Kohn
Chair, Mathematics Department
Princeton University

Welcome to the conference on Prospects in Mathematics which is being held as part of the celebration of the two hundred and fiftieth anniversary of the founding of Princeton University. It is a pleasure for me to greet you here on behalf of the Mathematics Department.

A couple of years ago, when plans were being drawn up for this celebration, all departments were asked how they would like to contribute. We had many discussions on this both in and out of department meetings. Somehow the department was guided by the glories of the past. In particular, most of us were thinking of the memorable picture of the participants in the mathematics conference which was held to celebrate the two hundredth anniversary of the founding of Princeton University. At that time it was still possible to have a conference in which a very large proportion of the leading mathematicians participated. I am particularly happy that in this conference we will have an historical link to the past given by Professors J. Milnor and D. C. Spencer.

Our discussions on how this conference should be organized gave rise to many diverse opinions; finally we agreed to set up a committee which would present the department with concrete plans. As you all know, committees can only work if they have an effective leader, and we were very fortunate that Professor Sinai agreed to become chair of this committee. He and his committee labored long and hard and I wish to take the opportunity to thank each of them. They are: W. Browder, S. Klainerman, P. Sarnak, and E. Stein. A conference like this also requires much administrative and logistical support. For this we are thankful to the staff of the department, and in particular the department manager Scott Kenney. We are also grateful to the Minerva Foundation, the National Science Foundation and Princeton University, each of which gave financial support to this enterprise. Finally, last but not least, I wish to convey the department's thanks to all participants.

Best wishes for a fruitful conference.

Growing Up in the Old Fine Hall

John Milnor

This talk will be antithetical to the main thrust of this conference. Rather than looking at the future, the focus will be on the past.

The title for this talk is meant quite literally: I was seventeen years old when I first came to Princeton in 1948 (just two years after the Princeton Bicentennial Conference on "Problems of Mathematics"). Fine Hall was in a real sense my home for many years. I was young and shy, and had no talent for dealing with people. Fine Hall was a wonderful new world for me. There was a common room where one could make oneself at home, and a marvelous library. Both students and faculty were friendly. Among the graduate students I particularly remember Serge Lang (who was very outspoken even then), and John Nash. I had never really heard much or reacted much to classical music before I came to Princeton. I first learned about Bach by listening to Nash wandering around the halls whistling. I also learned about Game Theory, and Nash equilibrium Theory. (Since I was very young, I assumed that it was the most natural thing in the world for a graduate student to develop a mathematical theory which changed the way that we think about social sciences. Compare [14].)

I especially enjoyed the games that were played in the Fine Hall common room. I learned the game of Go, not very well, but well enough to appreciate its subtlety and highly geometric strategy. There were games of Kriegspiel: chess played back to back on two different boards, with a referee in the middle keeping track of the position on a third chess board which neither player could see. Unfortunately, it happened often that the referee would make a mistake; then we said that the referee had "won" the game. Also there was the game which we called "Nash", introduced by John Nash (these days it is usually called "Hex"), a wonderful topological game of connections [14]. I was very fond of these things and felt very much at home.

Of course the most important thing about a department is its faculty. I don't have a photograph from this time, but Plate 1 shows the faculty 10 years later, when I was a fledgling Associate Professor. On the left is Don Spencer, who is with us today. The first person I met in Princeton was Al Tucker (front center). In fact I met him the summer before I enrolled in Princeton, when he interviewed me as a prospective freshman. As a freshman, I attended Tucker's differential geometry class. Here he posed a question which was my first real introduction to mathematics. Tucker described the concept of *curvature* for a closed space curve C with arc length s and with position vector $\mathbf{x} = \mathbf{x}(s)$. By definition, the curvature

---

1991 *Mathematics Subject Classification.* 01A65, 57M50.

PLATE 1. The Fine Hall Faculty in May 1958. Back: Spencer,
Church, Artin, Wilks, Milnor, Tukey, Steenrod, Feller. Middle:
Hirzebruch, Fox, Moore. Front: Wigner, Tucker, Bargmann.

is the norm of the second derivative vector

$$\kappa = \left\| \frac{d^2 \mathbf{x}}{ds^2} \right\|,$$

and the *total curvature* is the integral $\oint \kappa \, ds$ around the curve. Tucker described
the theorem of Fenchel which says that

$$\oint \kappa \, ds \geq 2\pi,$$

with equality if and only if C is a convex plane curve. Then he posed the problem of
Borsuk: *If the curve is knotted, does it follow that the total curvature must satisfy*

$$\oint \kappa \, ds > 4\pi?$$

PLATE 2. Solomon Lefschetz

A few days later, I succeeded in providing a proof. (A different proof by Istvan Fáry was published somewhat earlier than mine. See [2], [8], [11].)

Let me go back to Plate 1. The person who was closest to me in the early years was Ralph Fox. He was my advisor for my Senior thesis and also for my Doctoral thesis [12]. Fox was a wonderful teacher in a strange way. I believe that his lectures were often somewhat confused: he didn't always get things exactly right. The effect was that the class had to help him. (I am afraid that I have often unintentionally followed his example.) However we learned a great deal in this way. I particularly enjoyed the course in point set topology which he taught by a form of the R. L. Moore method: He told us the theorems and we had to produce the proofs. I can't think of a better way of learning how to make proofs and how to learn the basic facts of topology—it was a marvelous education.

Let me mention another teacher: Emil Artin, who was an extremely impressive and charismatic figure. His style was the exact opposite: each lecture was a totally polished work of art. There was not a misplaced epsilon, and everything was exactly efficient and perfectly done. In theory, this should be much more conducive to learning but in practice I am not sure that it was. For example, I attended a course in algebraic number theory from Artin which was extremely elegant, although perhaps too advanced for me. However, it wasn't until a few years later that I learned what an algebraic number was. The course was so streamlined that algebraic numbers were never actually mentioned.

Later the person who influenced me most was Norman Steenrod. The ideas and methods which I learned from him played a definitive role in my mathematical life. Another important influence was John Moore.

Let me go on to a couple of people who are not in Plate 1. If Ralph Fox was my mathematical father, then Solomon Lefschetz was my mathematical grandfather. Lefschetz was an amazing figure—I don't know how else to put it. He had to work with very clumsy artificial hands, which could hold a piece of chalk but not much else. Don Spencer tells us that his wife called him Sol, but I never heard anyone call him by his first name. He looks rather imposing in this picture, and he was imposing in person. It was not that he intended to be: he was very friendly and helpful. I remember that he would fix me with this gaze and give me all sorts of wise

PLATE 3. Salomon Bochner

and useful advice, but I didn't know quite how to take it. In fact I am afraid that sometimes I had to work hard to keep a straight face. I admired him greatly: I think he spoke all known languages, at least all languages I'd ever heard of, although as far as I could tell he had a strong accent in every one of them. And he seemed to know every part of mathematics—he had worked in algebraic geometry, topology, and the qualitative theory of differential equations (what we now call dynamics). He wrote books on many different subjects, and I am afraid that he was confusing on every subject, but his work was nevertheless useful and influential. Some years ago Don Davis at Lehigh tried to make a family tree for Lefschetz's mathematical descendents: his students, their students, and so on. After a few months' work he had to give up: there were just too many, so that it was impossible to fit all of them on a reasonable piece of paper. It was a phenomenal example of exponential growth. The descendants of Lefschetz have had, and continue to have, a great impact in many different branches of pure and applied mathematics.

One other person who doesn't appear in the group photograph is Salomon Bochner. I'm afraid that I never got to know Bochner at all well until many years later. In fact I really didn't learn much analysis during my years as a student. A few years after I graduated, when I was assigned to teach a course in second year Calculus, I was chagrined to discover that I didn't know much of the material and had to learn it as I went along. I had one experience with Bochner that I am quite ashamed of. It was the custom, as I understood it, to attend many different courses on the first day to try to decide which ones to go to. I went to Bochner's course in this way, once during my first years at Princeton, and decided after a few minutes that it really wasn't for me. (I was fascinated by geometry and topology but had no use for analysis.) However, Bochner picked me out, and assigned me to give a report. I didn't know how to deal with this, and I'm afraid that I didn't deal with it at all well—I simply disappeared and didn't come back. I'm sure that he forgave me eventually, but I didn't get to know him until many years later.

As Don mentioned, Lefschetz and Bochner didn't get along at all. I have never learned the reason, but they completely avoided each other. Yet there were many similarities: for example when they retired from Princeton, instead of giving up, each of them simply moved elsewhere and started teaching new groups of students.

Lefschetz moved to Mexico where he had an important influence, while Bochner moved to Rice University where he also had an important influence.

Now let me skip ahead a few years and talk about my favorite period of activity in Princeton. In the fifties I was very much interested in the fundamental problem of understanding the topology of higher dimensional manifolds. In particular, I focussed on the class of $2n$-dimensional manifolds which are $(n-1)$-connected, since these seemed like the simplest examples for which one had a reasonable hope of progress. (Of course the class of manifolds with the homotopy type of a sphere is even simpler. However the generalized Poincaré problem of understanding such manifolds seemed much too difficult: I had no idea how to get started.) For a closed $2n$-dimensional manifold M^{2n} with no homotopy groups below the middle dimension, there was a considerable body of techniques and available results to work with. First, one could easily describe the homotopy type of such a manifold. It can be built up (up to homotopy type) by taking a union of finitely many n-spheres intersecting at a single point, and then attaching a $2n$-cell e^{2n} by a mapping of the boundary ∂e^{2n} to this union of spheres, so that

$$M^{2n} \simeq (S^n \vee \cdots \vee S^n) \cup_f e^{2n}.$$

Here the attaching map f represents a homotopy class in $\pi_{2n-1}(S^n \vee \cdots \vee S^n)$, a homotopy group that one can work with effectively, at least in low dimensions. Thus the homotopy theory of such manifolds is under control. We can understand this even better by looking at cohomology. The cohomology of such an M^{2n}, using integer coefficients, is infinite cyclic in dimension zero, free abelian in the middle dimension with one generator for each of the spheres, and is infinite cyclic in the top dimension where we have a cohomology class corresponding to this top dimensional cell; that is

$$H^0(M^{2n}) \cong \mathbf{Z}, \qquad H^n(M^{2n}) \cong \mathbf{Z} \oplus \cdots \oplus \mathbf{Z}, \qquad H^{2n}(M^{2n}) \cong \mathbf{Z}.$$

The attaching map f determines a cup product operation: To any two cohomology classes in the middle dimension we associate a top dimensional cohomology class, or in other words (if the manifold is oriented) an integer. This gives a bilinear pairing from $H^n \otimes H^n$ to the integers. This pairing is symmetric if n is even, skew-symmetric if n is odd, and always has determinant ± 1 by Poincaré duality. For n odd this pairing is an extremely simple algebraic object. However for n even such symmetric pairings, or equivalently quadratic forms over the integers, form a difficult subject which has been extensively studied. (See [15], and compare [10].) One basic invariant is the *signature*, computed by diagonalizing the quadratic form over the real numbers, and then taking the number of positive entries minus the number of negative entries.

So far this has been pure homotopy theory, but if the manifold has a differentiable structure, then we also have characteristic classes, in particular the Pontrjagin classes in dimensions divisible by four,

(1) $$p_i \in H^{4i}(M).$$

This was the setup for the manifolds that I was trying to understand as a long term project during the 50's. Let me try to describe the state of knowledge of topology in this period. A number of basic tools were available. I was very fortunate in learning about cohomology theory and the theory of fiber bundles from Norman Steenrod, who was a leader in this area. These two concepts are

combined in the theory of characteristic classes [16], which associates cohomology
classes in the base space to certain fiber bundles. Another basic tool is obstruction
theory, which gives cohomology classes with coefficients in appropriate homotopy
groups. However, this was a big sticking point in the early 50's because although one
knew very well how to work with cohomology, no one had any idea how to compute
homotopy groups except in special cases: most of them were totally unknown. The
first big breakthrough came with Serre's thesis, in which he developed an algebraic
machinery for understanding homotopy groups. A typical result of Serre's theory
was that the stable homotopy groups of spheres

$$\Pi_n = \pi_{n+k}(S^k) \quad (k > n + 1)$$

are always finite. Another breakthrough in the early 50's came with Thom's cobordism theory. Here the basic objects were groups whose elements were equivalence
classes of manifolds. He showed that these groups could be computed in terms
of homotopy groups of appropriate spaces. As an immediate consequence of his
work, Hirzebruch was able to prove a formula which he had conjectured relating
the characteristic classes of manifolds to the signature. For any closed oriented
$4m$-dimensional manifold, we can form the signature of the cup product pairing

$$H^{2m}(M^{4m}; \mathbf{R}) \otimes H^{2m}(M^{4m}; \mathbf{R}) \to H^{4m}(M^{4m}; \mathbf{R}) \cong \mathbf{R},$$

using real coefficients. If the manifold is differentiable, then it also has Pontrjagin
classes (1). Taking products of Pontrjagin classes going up to the top dimension
we build up various *Pontrjagin numbers*. These are integers which depend on the
structure of the tangent bundle. Hirzebruch conjectured a formula expressing the
signature as a rational linear combination of the Pontrjagin numbers. For example

(2) $$\text{signature}(M^4) = \tfrac{1}{3} p_1[M^4]$$

and

(3) $$\text{signature}(M^8) = \tfrac{1}{45}(7p_2 - (p_1)^2)[M^8].$$

Everything needed for the proof was contained in Thom's cobordism paper, which
treated these first two cases explicitly, and provided the machinery to prove Hirzebruch's more general formula.

   These were the tools which I was trying to use in understanding the structure
of $(n-1)$-connected manifolds of dimension $2n$. In the simplest case, where the
middle Betti number is zero, these constructions are not very helpful. However in
the next simplest case, with just one generator in the middle dimension and with
$n = 2m$ even, they provide quite a bit of structure. If we try to build up such a
manifold, as far as homotopy theory is concerned we must start with a single $2m$-
dimensional sphere and then attach a cell of dimension $4m$. The result is supposed
to be homotopy equivalent to a manifold of dimension $4m$:

$$S^{2m} \cup e^{4m} \simeq M^{4m}.$$

What can we say about such objects? There are certainly known examples; the
simplest is the complex projective plane in dimension four—we can think of that
as a 2-sphere (namely the complex projective line) with a 4-cell attached to it.

   Similarly in dimension eight there is the quaternionic projective plane which
we can think of as a 4-sphere with an 8-cell attached, and in dimension sixteen
there is the Cayley projective plane which has similar properties. (We have since
learned that such manifolds can exist only in these particular dimensions.)

Consider a smooth manifold M^{4m} which is assumed to have a homotopy type which can be described in this way. What can it be? We start with a $2m$-dimensional sphere S^{2m}, which is certainly well understood. According to Whitney, this sphere can be smoothly embedded as a subset $S^{2m} \subset M^{4m}$ generating the middle dimensional homology, at least if $m > 1$. We look at a tubular neighborhood of this embedded sphere, or equivalently at its normal $2m$-disk bundle E^{4m}. In general this must be twisted as we go around the sphere—it can't be simply a product or the manifold wouldn't have the right properties. In terms of fiber bundle theory, we can look at it in the following way: Cut the $2m$-sphere into two hemispheres D_+^{2m} and D_-^{2m}, intersecting along their common boundary S^{2m-1}. Over each of these hemispheres we must have a product bundle, and we must glue these two products together to form

$$E^{4m} = \left(D_+^{2m} \times D^{2m}\right) \cup_F \left(D_-^{2m} \times D^{2m}\right).$$

Here the gluing map $F(x,y) = (x, f(x)y)$ is determined by a mapping

$$f : S^{2m-1} \to \mathrm{SO}(2m)$$

from the intersection $D_+^{2m} \cap D_-^{2m}$ to the rotation group of D^{2m}. Thus the most general way of thickening the $2m$-sphere can be described by an element of the homotopy group $\pi_{2m-1}\mathrm{SO}(2m)$. In low dimensions, this group was well understood.

In the simplest case $4m = 4$, we start with a D^2-bundle over S^2 determined by an element of $\pi_1\mathrm{SO}(2) \cong \mathbf{Z}$. It is not hard to check that the only 4-manifold which can be obtained from such a bundle by gluing on a 4-cell is (up to orientation) the standard complex projective plane: This construction does not give anything new. The next case is much more interesting. In dimension eight we have a D^4-bundle over S^4 which is described by an element of $\pi_3(\mathrm{SO}(4))$. Up to a 2-fold covering, the group $\mathrm{SO}(4)$ is just a Cartesian product of two 3-dimensional spheres, so that $\pi_3\mathrm{SO}(4) \cong \mathbf{Z} \oplus \mathbf{Z}$. More explicitly, identify S^3 with the unit 3-sphere in the quaternions. We get one mapping from this 3-sphere to itself by left multiplying by an arbitrarily unit quaternion and another mapping by right multiplying by an arbitrary unit quaternion. Putting these two operations together, the most general $(f) \in \pi_3(\mathrm{SO}(4))$ is represented by the map $f(x)y = x^i y x^j$, where x and y are unit quaternions and where $(i, j) \in \mathbf{Z} \oplus \mathbf{Z}$ is an arbitrary pair of integers.

Thus to each pair of integers (i, j) we associate an explicit 4-disk bundle over the 4-sphere. We want this to be a tubular neighborhood in a closed 8-dimensional manifold, which means that we want to be able to attach a 8-dimensional cell which fits on so as to give a smooth manifold. For that to work, the boundary $M^7 = \partial E^8$ must be a 7-dimensional sphere S^7. The question now becomes this: For which i and j is this boundary isomorphic to S^7? It is not difficult to decide when it has the right homotopy type: In fact M^7 has the homotopy type of S^7 if and only if $i + j$ is equal to ± 1. To fix our ideas, suppose that $i + j = +1$. This still gives infinitely many choices of i. For each choice of i, note that $j = 1 - i$ is determined, and we get as boundary a manifold $M^7 = \partial E^8$ which is an S^3-bundle over S^4 having the homotopy type of S^7. Is this manifold S^7, or not?

Let us go back to the Hirzebruch–Thom signature formula (3) in dimension 8. It tells us that the signature of this hypothetical 8-manifold can be computed from $(p_1)^2$ and p_2. But the signature has to be ± 1 (remember that the quadratic form always has determinant ± 1), and we can choose the orientation so that it is $+1$. Since the restriction homomorphism maps $H^4(M^8)$ isomorphically onto

$H^4(S^4)$, the Pontrjagin class p_1 is completely determined by the tangent bundle in a neighborhood of the 4-sphere, and hence by the integers i and j. In fact it turns out that p_1 is equal to $2(i - j) = 2(2i - 1)$ times a generator of $H^4(M^8)$, so that $p_1^2[M^8] = 4(2i - 1)^2$. We have no direct way of computing p_2, which depends on the whole manifold. However, we can solve equation (3) for $p_2[M^8]$, to obtain the formula

$$(4) \qquad p_2[M^8] = \frac{p_1^2[M^8] + 45}{7} = \frac{4(2i - 1)^2 + 45}{7}.$$

For $i = 1$ this yields $p_2[M^8] = 7$, which is the correct answer for the quaternion projective plane. But for $i = 2$ we get $p_2[M^8] = \frac{81}{7}$, which is impossible! Since p_2 is a cohomology class with integer coefficients, this Pontrjagin number $p_2[M^8]$, whatever it is, must be an integer.

What can be wrong? If we choose p_1 in such a way that (4) does not give an integer value for $p_2[M^8]$, then there can be no such differentiable manifold. The manifold $M^7 = \partial E^8$ certainly exists and has the homotopy type of a 7-sphere, yet we cannot glue an 8-cell onto E^8 so as to obtain a smooth manifold. What I believed at this point was that such an M^7 must be a counterexample to the seven dimensional Poincaré hypothesis: I thought that M^7, which has the homotopy type of a 7-sphere, could not be homeomorphic to the standard 7-sphere.

Then I investigated further and looked at the detailed geometry of M^7. This manifold is a fairly simple object: an S^3-bundle over S^4 constructed in an explicit way using quaternionic multiplication. I found that I could actually prove that it was homeomorphic to the standard 7-sphere, which made the situation seem even worse! On M^7, I could find a smooth real-valued function which had just two critical points: a non-degenerate maximum point and a non-degenerate minimum point. The level sets for this function are 6-dimensional spheres, and by deforming in the normal direction we obtain a homeomorphism between this manifold and the standard S^7. (This is a theorem of Reeb: if a closed k-manifold possesses a Morse function with only two critical points, then it must be homeomorphic to the k-sphere.) At this point it became clear that what I had was not a counterexample to the Poincaré hypothesis as I had thought. This M^7 really was a topological sphere, but with a strange differentiable structure.

There was a further surprising conclusion. Suppose that we cut this manifold open along one of the level sets, so that

$$M^7 = D_+^7 \cup_f D_-^7,$$

where the $D_\pm^7$ are diffeomorphic to 7-disks . These are glued together along their boundaries by some diffeomorphism $g : S^6 \to S^6$. *Thus this manifold M^7 can be constructed by taking two 7-dimensional disks and gluing the boundaries together by a diffeomorphism.* Therefore, at the same time, the proof showed that there is a diffeomorphism from S^6 to itself which is essentially exotic: It cannot be deformed to the identity by a smooth isotopy, because if it could then M^7 would be diffeomorphic to the standard 7-sphere, contradicting the argument above.[1]

This talk has been about the past, but perhaps I should close by saying something with a bearing on the future, something about my philosophy of mathematics.

---

[1]For accounts of exotic spheres, see [5]–[7], [9], [13]. The classification problem for $(n - 1)$-connected $2n$-dimensional manifolds was finally completed by Wall [18], making use of exotic spheres.

What I love most about the study of mathematics is its anarchy! There is no mathematical czar who tells us which direction we must work in, what we must be doing. There are thousands of mathematicians all over the world each going in his or her own direction. Many are exploring the most popular or fashionable directions, but others work in strange or unfashionable directions. Perhaps many are going the wrong way, but cumulatively the many different directions, the many different approaches, mean that new and often unexpected things will be discovered. I like to picture the frontier of mathematics as a great ragged wall, with the unknown, the unsolved problems, to one side, and with thousands of mathematicians on the other side, each trying to nibble away at different parts of the problem using different approaches. Perhaps most of them don't get very far, but every now and then one of them breaks through and opens a new area of understanding. Then perhaps another one makes another breakthrough and opens another new area. Sometimes these breakthroughs come together, so that we have different parts of mathematics merging, giving us wide new perspectives. Often the people who make these breakthroughs are those who are well known, those we expect to obtain good results; but not always. Many times major results are obtained by those who are not at all well known, or by people we may know but underestimate, so that we are completely surprised to find that they have accomplished so much. It is wonderful that no one has the power to turn such people off! Of course they can be discouraged, and often have to fight for recognition, but there are many universities, many places where one can do mathematical research, and no astronomical budget is required. Thus there is always hope that even people who have unpopular ideas will have a real chance to succeed.

Here is one example from my early days in Fine Hall: Papakyriakopoulos was certainly not a very conspicuous member of the Princeton mathematical community. He worked steadily here for many years, apparently not accomplishing very much. Then suddenly he made a breakthrough with his proof of Dehn's lemma, and shortly after with his proof of the sphere theorem [**17**]. This was the beginning of a whole new era in three dimensional topology.

I am very grateful that I never had the job of deciding which directions should be pursued in mathematics. Let me describe the way that I felt about topology back in the fifties. I thought that the good things to study were smooth manifolds and well behaved cell complexes; the good methods were methods from algebraic topology and differential geometry. I don't mean to say now that these are not good things to study: I love them still and they are extremely important. But at that time I was completely uninterested in other parts of topology, for example the study of complicated decompositions of 3-space, or infinite dimensional spaces, or nasty sets like indecomposable continua. I thought that these were totally boring and not worth studying. Yet later some of the basic problems in which I was very much interested came to be solved by such methods. For example, I worried a great deal about the problem of the topological invariance of Whitehead and Reidemeister torsion [**12**]. These are invariants for a complex with non-trivial fundamental group which are computed by passing to the universal covering and studying the action of the fundamental group on the covering cell complex. This is a very combinatorial object and it is not at all clear that you can make sense of it topologically. The problem was solved by Chapman [**1**] who treated it by going completely outside of the kind of mathematics I was interested in. He worked with infinite dimensional manifolds modeled on the Hilbert cube I^∞ (an infinite Cartesian product of closed unit intervals), and proved a beautiful theorem: Given two finite complexes, there is a homotopy equivalence between them with zero Whitehead torsion if and only if they become homeomorphic after multiplying both by I^∞. As an immediate corollary, homeomorphic complexes have the same simple homotopy type: there is always a map between them with zero Whitehead torsion.

Another example is the four dimensional topological Poincaré conjecture. This is not a problem that I ever seriously tried to work on, but it was clearly one of the basic problems in topology. It was solved by Mike Freedman using methods from the Bing school of topology, exactly the kind of topology which I had always avoided.

Here is another example. I mentioned indecomposable continua as objects which seemed especially unpleasant to me. However, in recent years I have been working in dynamics. Thus I am interested in understanding the global behavior of differential equations (smooth differential equations in euclidean space or on a smooth manifold), and want to understand the long term behavior of solutions. The startling fact that has emerged over the past 10 or 20 years is that often wild objects like indecomposable continua are exactly what are needed in order to understand this behavior [4]. Even for smooth and natural differential equations, the limit sets can be these complicated objects. To understand the behavior of the solutions, the things with real physical meaning, one has to study the fractal geometry associated with these strange sets.

I am making a plea for mathematical tolerance: Even if some branch of mathematics seems uninteresting today, it should not be given up completely. It is important to have people working in many different directions with many different points of view in order to attack the basic problem of understanding the mathematical world, and its applications.

References

[1] T. A. Chapman, *Topological invariance of Whitehead torsion*, Amer. J. Math. **96** (1974) 488–497.

[2] I. Fáry, *Sur la courbure totale des courbes fermées*, Annales Soc. Polonaise **20** (1947) 251–265.

[3] M. H. Freedman, *The topology of four-dimensional manifolds*, J. Diff. Geom. **17** (1982) 357–453.

[4] J. Kennedy, *How indecomposable continua arise in dynamical systems*, Papers on general topology and applications (Madison WI 1991) 180–201; Ann. New York Acad. Sci. **704** 1993.

[5] M. Kervaire and J. Milnor, *Groups of homotopy spheres I*, Annals Math. **77**, 504–537 (1963).

[6] A. Kosinski, "Differential Manifolds", Academic Press (1993).

[7] T. Lance, *Differentiable structures on manifolds*, to appear in "Surgery theory and its applications: surveys presented in honor of C. T. C. Wall", edited by J. Rosenberg et al.

[8] J. Milnor, *On the total curvature of knots*, Annals Math. **52** (1950) 248–257.

[9] ———, *On manifolds homeomorphic to the 7-sphere*, Annals Math. **64**, 399–405 (1956).

[10] ———, *On simply connected 4-manifolds*, 122–128 of "Symposium Internacional de Topologia Algebraica", UNAM and UNESCO, Mexico (1958).

[11] ———, "Collected Papers 1, Geometry", Publish or Perish 1994.

[12] ———, "Collected Papers 2, The Fundamental Group", Publish or Perish 1995.

[13] ———, "Collected Papers 3, Differential Topology", Publish or Perish (in preparation).

[14] ———, *A Nobel prize for John Nash*, Math. Intelligencer **17** (1995) 11–17.

[15] ——— and D. Husemoller, "Symmetric Bilinear Forms", Springer (1973).

[16] ——— and J. Stasheff, "Characteristic Classes", Ann. Math. Stud. 76, Princeton (1974).

[17] C. Papakyriakopoulos, *On Dehn's lemma and the asphericity of knots*, Annals of Mathematics **66** (1957).

[18] C. T. C. Wall, *Classification of $(n-1)$-connected $2n$-manifolds*, Annals Math. **75**, 163–189 (1962).

SUNY AT STONY BROOK, INSTITUTE FOR MATHEMATICAL SCIENCES, MATHEMATICS BUILDING, STONY BROOK, NY 11794-3660

E-mail address: jack@math.sunysb.edu

Old Memories and an Old Problem

Donald Spencer

I. Reminiscences

1. I am honored and happy to be here. This department, recently named by the National Research Council as the best in the U.S., has a brilliant history beginning shortly after the turn of the century with the appointment of Eisenhart, Veblen, James Jeans and others. But the past cannot outshine the last few years, during which Wiles proved Fermat's Last Theorem and Nash received a Nobel Prize for his doctoral thesis, not to mention the recent techniques and results to be found in the works of Kohn and Fefferman and in the work of Stein. I apologize for neglecting to mention other brilliant accomplishments in the department, which are unknown to me or further removed from my field, such as the contribution of Shimura stated by Wiles in the introduction of his paper.

2. I arrived in Princeton in September, 1949, as visiting professor from Stanford after completing an AMS Colloquium Volume with Schaeffer on the coefficient domains of univalent (schlicht) functions. A remnant of this research hangs on the wall to the left of the entrance to Fine Hall; it is a model of a 3-dimensional section of the domain of the first two coefficients of a univalent function. The following year (1950) I accepted a non-tenured associate professorship at Princeton University, a step down from Stanford, and became a professor here in 1953.

3. The department in 1949–50. I was overwhelmingly impressed by the department's faculty and students. The department was then located in the old Fine Hall, a beautiful building now called Jones Hall, the first two floors of which were built around a central auditorium. Professors in the department then were Artin, Bochner, Church, Lefschetz (chairman), Tucker, Wigner and Wilks, associate professors Bargmann, Fox, Steenrod and Tukey, and instructors Brownell, Gale, Kuhn and MacKenzie. Among the assistants were Ankeny, Calabi, Kemeny, McCarthy, Sampson. Nash, Washnitzer, Hawley and Lang were some of the graduate students. And, in the fall of 1948, Jack Milnor entered the department as a freshman; shortly thereafter he solved an important problem in knot theory, the beginning of his brilliant career. Agnes Fleming (Henry) was the main secretary, a remarkable woman whom Lefschetz had earlier brought to the department when she was eighteen.

1991 *Mathematics Subject Classification.* 01A65, 11L03.

4. The daily tea. Fine Hall was a remarkable community whose members attended a tea starting at 4:00 in the Common Room. All of us knew and talked to each other and the atmosphere was usually warm and friendly. The graduate students would sometimes talk about the faculty, characterizing them in succinct and humorous ways. I recall a few of these characterizations, namely: Lefschetz never stated a false theorem or gave a correct proof; Bochner says one thing, writes another and means a third; Church is so cautious that, when he signs his name, he counts the letters in it (Agnes). Bochner and Lefschetz did not get along with each other personally but each admired the other as mathematician. Bochner said that Lefschetz was the greatest living homologist and Lefschetz said "damn it, Bochner is a fine mathematician". Nevertheless, the department on the whole was a very friendly one.

5. Kodaira became a member of the Institute for Advanced Study in the fall of 1949, having been brought there by Weyl who was impressed by Kodaira's paper entitled "Harmonic fields in riemannian manifolds (generalized potential theory)", Annals of Math. **50** (1949), 587–665. I met Kodaira almost immediately after his arrival from Japan, and invited him to lecture on this paper at the University, which he did. Thus began our collaboration which lasted almost 18 years.

6. The University and the Institute in the 1950's. These two institutions created in this decade what Atiyah, in a recent article, called a golden age. I agree, but lack of time prevents me from reminiscing further. To do so would require a series of lectures.

7. A worry about the future. I am concerned that the explosive exponential growth of the world's human population will bring so much stress to our daily lives that time for contemplation and creative thinking will be diminished to the point where scientific research will suffer and slow down. Some of my friends believe that telecommunication will mitigate many of the effects of high population density, for example by enabling people to be educated and to work in their homes. I hope they are right. Finally big is not better in the university. Princeton has been smart in trying to curtail growth, putting quality above size.

II. An unsolved problem, known as Littlewood's Problem

In my Cambridge thesis, written under J. E. Littlewood in the years immediately preceding the outbreak in Europe of World War II, a problem in non-linear Diophantine approximation arose which, in the simplest case, can be stated as follows.

If y is a real number, denote by $\|y\|$ the distance from y to the nearest integer. Then, does there exist a pair (θ_1, θ_2) of real numbers, necessarily irrational and of bounded partial quotients in their continued fraction developments, such that

$$\|n\theta_1\| \cdot \|n\theta_2\| \geq \frac{K}{n}, \quad \text{for } n = 1, 2, 3, \ldots,$$

where $K = K(\theta_1, \theta_2)$ is a positive number? If no such pairs exist, what is the positive, monotonically increasing function $\phi(n)$ of essentially slowest growth such that

$$\|n\theta_1\| \cdot \|n\theta_2\| \geq \frac{K}{n\phi(n)}, \quad n = 1, 2, 3, \ldots, \tag{1}$$

holds for some $(\theta_1, \theta_2) \in \mathbb{R}^2$? For example, can we have $\phi(n) = \log n$ or $\log^2 n$ for $n \geq 2$?

My thesis was published in two parts, the first part is [5] and the second is [6]. It is the second which is relevant here and, in particular, there it is shown that for each $\varepsilon > 0$, (1) holds for almost all $(\theta_1, \theta_2) \in \mathbb{R}^2$ with $\phi(n) = \log^{2+\varepsilon} n$ for $n \geq 2$. More generally, for almost all $(\theta_1, \theta_2, \ldots, \theta_k) \in \mathbb{R}^k$, $k \geq 1$, we have, for each $\varepsilon > 0$,

$$\lim_{n \to \infty} n \log^{k+\varepsilon} n \cdot \|n\theta_1\| \cdots \|n\theta_k\| = \infty.$$

This result has been shown to be best possible by Gallagher [1], who proves that

$$\varliminf n \log^k n \|n\theta_1\| \cdots \|n\theta_k\| = 0$$

almost everywhere in $\mathbb{R}^k$. And, of course, for any $k \geq 2$, we can ask for the function $\phi_k(n)$ of essentially slowest growth such that there exists a point $(\theta_1, \theta_2, \ldots, \theta_k) \in \mathbb{R}^k$ with

$$\|n\theta_1\| \cdots \|n\theta_k\| \geq \frac{K_k}{n\phi_k(n)}.$$

This problem arose naturally in [6], as we shall now indicate briefly.

Hardy and Littlewood, in a series of papers ([2], [3] and [4]), examined in fine detail the asymptotic behavior of the number of lattice points in a right-angled triangle. As my thesis problem, Littlewood suggested the generalization of this study to the case of multi-dimensional tetrahedra. Suppose that $\eta, \omega_1, \omega_2, \ldots, \omega_s$, where $s \geq 2$, are positive real numbers, and let $N^{(s)}(\eta)$ be the sum over all $m_k \geq 1$ ($k = 1, 2, \ldots, s$) for which $m_1\omega_1 + m_2\omega_2 + \cdots + m_s\omega_s \leq \eta$. Then $N^{(s)}(\eta)$ is the number of lattice points lying inside or on the boundary of the s-dimensional tetrahedron in $\mathbb{R}^s$ bounded by the s coordinate hyperplanes $x_1 = 1$, $x_2 = 1$, $\ldots$, $x_s = 1$ and the hyperplane $\omega_1 x_1 + \omega_2 x_2 + \cdots + \omega_s x_s = \eta$. We have (see [6])

$$N^{(s)}(\eta) = R^{(s)}(\eta) + T^{(s)}(\eta)$$

where $R^{(s)}(\eta)$ is a polynomial in η of degree s whose coefficients are symmetric functions of $\omega_1, \omega_2, \ldots, \omega_s$ and $T^{(s)}(\eta)$ is the error term. We assume that all ratios of distinct ω's are irrational; in that case it is the asymptotic behavior of $T^{(s)}(\eta)$, as η tends to infinity, which is of interest. It is shown in [6] that

$$T^{(s)}(\eta) = \frac{1}{2^{s-1}\pi} \sum_{j=1}^{s} \sum_{\lambda_j=1}^{\infty} \frac{\cos\left(\frac{2\lambda_j \pi}{\omega_j}\left(\eta - \frac{1}{2}\sum_{i \neq j}\omega_i\right) - \frac{s\pi}{2}\right)}{\lambda_j \prod_{\substack{k=1 \\ k \neq j}}^{s} \sin\frac{\lambda_j \omega_k \pi}{\omega_j}} \tag{2}$$

where the summation is effected as follows: the partial sums are formed of those terms of the s separate series for which $\lambda_j < \omega_j Y / 2\pi$ ($j = 1, 2, \ldots, s$), and the limit of the sum of these series is taken as $Y \to \infty$ through a sequence all of whose members differ by a fixed amount $L(\omega_1, \omega_2, \ldots, \omega_s)$ from any numbers of the forms $2m_j \pi / \omega_j$, for m_j an integer. After Hardy, (2) is called a compound series.

The easiest result concerning the asymptotic behavior of $T^{(s)}(\eta)$ is the metric one, namely (see [6]): for almost all $(\omega_1, \omega_2, \ldots, \omega_s) \in \mathbb{R}^s$ and any $\varepsilon > 0$, we have

$$T^{(s)}(\eta) = O\{(\log \eta)^{s+\varepsilon}\}. \tag{3}$$

The proof of (3) is based on the following lemma: the series

$$\sum_{m=2}^{\infty} \frac{1}{m \log^{k+1+\varepsilon} m \prod_{j=1}^{k} |\sin m\pi\theta_j|}$$

is convergent for each $\varepsilon > 0$ and almost all $(\theta_1, \theta_2, \ldots, \theta_k) \in \mathbb{R}^k$, $k \geq 1$. From this we have as an immediate consequence, for almost all $(\theta_1, \theta_2, \ldots, \theta_k)$,

$$\sum_{m=1}^{n} \frac{1}{m \prod_{j=1}^{k} |\sin m\pi\theta_j|} = O(\log^{k+1+\varepsilon} n).$$

But, in the case $k \geq 2$, deeper results are apparently not known, and this area of non-linear Diophantine approximation seems to be wide open for further research.

References

[1] P. Gallagher, "Metric simultaneous diophantine approximation", Journal London Math. Soc. **37** (1962), 387–390.

[2] G. H. Hardy and J. E. Littlewood, "Some problems of diophantine approximation", Proc. 5th Int. Congress of Mathematicians 1912, 223–229.

[3] G. H. Hardy and J. E. Littlewood, "The lattice points of a right-angled triangle", Proc. London Math. Soc. (2) **20** (1921), 15–36.

[4] G. H. Hardy and J. E. Littlewood, "The lattice points of a right-angled triangle (second memoir)", Hamburg Math. Abhandlungen **1** (1922), 212–249.

[5] D. C. Spencer, "On a Hardy–Littlewood problem of diophantine approximation", Proc. Cambridge Philosophical Soc. **35** (1939), 527–547.

[6] D. C. Spencer, "The lattice points of tetrahedra", Journ. Math. and Phys. **31**:3 (October 1942), 24–32.

PROFESSOR EMERITUS, MATHEMATICS DEPARTMENT, PRINCETON UNIVERSITY, PRINCETON, NJ 08540, UNITED STATES

E-mail address: To be supplied

The Electron Is Inexhaustible*

Jürg Fröhlich

It's not philosophy we are after, but the behavior of real things.

– R. P. Feynman

1991 *Mathematics Subject Classification.* 35Q40, 53C80, 81T25.

This paper is dedicated to the memory of my colleague and friend, the late Claude Itzykson.

*V. I. Ulyanov, alias Lenin.

1. Introduction

Two hundred and fifty years ago, on January 12, 1746, Johann Heinrich Pestalozzi was born in Zürich. He was to be a great benefactor of poor children. As importantly, he would be remembered as the founder of public schools in Europe open to children from any social background and free of charge.[1]

Among the fairly numerous Pestalozzis in my life I should like to mention, besides my former teachers and supporters at ETH–Zürich, Barry Simon and Arthur Wightman, who convinced the mathematics department of Princeton University to offer me an assistant professorship at a rather crucial point in my career when I was a young and fairly poorly educated man. Not only did Princeton admit me free of charge, but the University paid my services generously enough that I was able to live happily and support a small family. Whether the University's investment in my own career was the right thing to do, at that time, or not, may remain an open question. But, in average, investing in careers of young scientists will pay back. It may be useful for politicians, educators and scientists to return to the study of some of the ideas and ideals of Pestalozzi.

Fifty years ago, Princeton was celebrating its bicentennial (and, incidentally, I was born). Among outside guests present at the celebration of the physics department were people like N. Bohr, P. A. M. Dirac, R. P. Feynman and numerous further distinguished physicists. Some of them were trying to understand and solve the divergence problems in quantum electrodynamics; but, at that time, they were stuck. Measurements of the Lamb shift and the Shelter Island Conference would come a year later. The bicentennial took place just shortly before physics was to enter one of its golden ages. Times were different, fifty years ago, perhaps more promising for theoretical physicists than now. Nowadays, we again appear to be a little stuck. Some of the great open problems, such as unifying quantum theory with the theory of gravitation, or understanding physical properties of strongly correlated (strongly coupled) systems with infinitely many degrees of freedom, such as low-energy QCD or high-temperature superconductors, remain elusive, in spite of significant progress.

The purpose of my lecture is to pursue some of the themes that must have interested people, fifty and more years ago, and, yet, remain interesting. More specifically, I shall argue that non-relativistic quantum theory remains a gold mine for mathematical inspiration—in fact, the mathematics inspired by quantum theory may well go far beyond that inspired by classical physics—and I shall suggest that we are far from understanding non-relativistic quantum theory really well.

The best understood quantum mechanical particles are the *photons*, i.e., the quanta of light (brought to light by Einstein) and the *non-relativistic, spinning electrons* (as first understood by Pauli). They are the main characters of my lecture. I should like to explore what they could have taught us and might still teach us about mathematics.

---

[1]For readers, who suspect that I might be one of those complacent Swiss, I should like to add that, among my favorite compatriots, there are the two well known writers and play wrights, the late Friedrich Dürrenmatt and the late Max Frisch. Those readers who are familiar e.g. with their plays "The Visit", "Frank V", "The 'honest' man and the incendiaries", "Andorra", know that some citizens of that little country, long favored by its fate in the shadow of European history, saw some of the weed that had grown in the shadow and tried to expose it. But that is another matter.

Before we get started on discussing electrons and photons, I should like to seize the opportunity and thank the organizers of this conference for the honor and the pleasure to celebrate with you the two hundred and fifty years of the College of New Jersey and of Princeton University, and my friend Tom Spencer, who played a particularly important role in my scientific life, for hospitality at the Institute for Advanced Study.

To conclude this introduction, I should warn the mathematicians (among potential readers of these notes) that they will encounter quite a number of unfamiliar looking notions and concepts and references to facts of physics. I hope they will cope with this circumstance in a positive spirit and take it as an encouragement to learn some more theoretical and mathematical physics and use them as a source of inspiration for interesting mathematical problems.

Obviously, these notes are not written in the style of a mathematics paper. However, I trust that wherever I try to give the impression that there is some real mathematics hiding behind some loose (and, perhaps, confusing) discussion there are, indeed, precise mathematical results that can be found in the literature referred to in the next. I *am* convinced that the topics discussed in these notes have led and will lead to plenty of interesting mathematics.

2. Differential geometry and topology from Pauli's quantum theory of the electron and of positronium

In non-relativistic quantum theory, time is a real parameter and space is some smooth, Riemannian spin^c manifold (M, g). Although, usually, (M, g) is three-dimensional Euclidian space, it is interesting to imagine that M is a general, n-dimensional (compact) spin^c manifold. Let $g = (g_{ij})$ be a Riemannian metric on the tangent bundle TM, and let $G = (g^{ij})$ denote the corresponding inverse metric on the cotangent bundle T^*M. By $\Lambda^{\boldsymbol{\cdot}} M$ we denote the bundle of completely antisymmetric, covariant tensors over M; the space of smooth sections, $\Omega^{\boldsymbol{\cdot}}(M)$, of $\Lambda^{\boldsymbol{\cdot}} M$ consists of differential forms on M. Let ∇ denote the Levi-Civita connection on $\Lambda^{\boldsymbol{\cdot}} M$. Given a vector bundle E over M, let $\Gamma(E)$ denote the space of smooth sections of E which is a projective module for the algebra

$$(2.1) \qquad\qquad \mathcal{A} = C^\infty(M)$$

of smooth functions on M. The bundle E is trivial if and only if $\Gamma(E)$ is a free $\mathcal{A}$-module. Important examples of vector bundles over M are $\Lambda^{\boldsymbol{\cdot}} M$, $\mathrm{Cl}(M)$ (the *Clifford bundle*), and—since M is assumed to be spin^c—the spinor bundle, S, and the conjugate spinor bundle, $\overline{S}$. The fibres, $\mathrm{Cl}(T_x^*M) \simeq \mathrm{Cl}_n$, of $\mathrm{Cl}(M)$ act irreducibly on the fibres, $\overset{(-)}{S}_x \simeq \mathbb{C}^{2^{[\frac{n}{2}]}}$, of $\overset{(-)}{S}$. If $\{\varepsilon^1, \ldots, \varepsilon^n\}$ is a basis of T_x^*M then $\{c(\varepsilon^1), \ldots, c(\varepsilon^n)\}$ are generators of $\mathrm{Cl}(T_x^*M)$ satisfying

$$(2.2) \qquad \{c(\varepsilon^i), \, c(\varepsilon^j)\} = -2G_x(\varepsilon^i, \varepsilon^j) = -2g^{ij}(x),$$

(where $\{A, B\} := AB + BA$, for linear operators A and B). If $\{\varepsilon^1, \ldots, \varepsilon^n\}$ is an orthonormal basis of T_x^*M we define

$$(2.3) \qquad\qquad \gamma_x := i^{[\frac{n+1}{2}]} \, c\left(\varepsilon^1\right) \ldots c\left(\varepsilon^n\right).$$

The elements $\gamma_x \in \mathrm{Cl}(T_x^*M), x \in M$, determine a *global involution*, γ, of $\Gamma(\mathrm{Cl}(M))$, because M is orientable. If n is odd γ is central, while if n is even $\{\gamma, c(\varepsilon^j)\} = 0$, for all j.

A connection ∇^S on S is called a spin^c connection if and only if it satisfies the "Leibniz rule"

$$(2.4) \qquad\qquad \nabla_X^S(c(\xi)\psi) = c(\nabla_X \xi)\psi + c(\xi)\nabla_X^S \psi,$$

where X is an arbitrary vector field, ξ an arbitrary one-form, and ψ an arbitrary spinor (section of $\Gamma(S)$) on M.[2]

If ∇_1^S and ∇_2^S are two hermitian spin^c connections on S then

$$\left(\nabla_1^S - \nabla_2^S\right)\psi = i\alpha \otimes \psi,$$

for arbitrary $\psi \in \Gamma(S)$, where α is a real one-form. The physical interpretation of α is that it is the difference of two *electromagnetic vector potentials* ("virtual U(1)-connections on S"). If R_{∇^S} denotes the curvature of a spin^c connection ∇^S then

$$(2.5) \qquad\qquad 2^{-\left[\frac{n}{2}\right]}\,\mathrm{Trace}\left(R_{\nabla^S}(X,Y)\right) = F_{\frac{1}{2}A}(X,Y),$$

for arbitrary vector fields X and Y, where F_A is the curvature ("electromagnetic field strength") of a U(1)-connection A ("vector potential") on a line bundle canonically associated to $S \otimes S$.

The Pauli-Dirac operator associated with a spin^c connection ∇^S on S is defined by

$$(2.6) \qquad\qquad D \equiv D_A = c \circ \nabla^S,$$

which acts on $\Gamma(S)$. To ∇^S there corresponds a unique (complex) conjugate connection $\overline{\nabla}^S$ on $\bar{S}$, and we define

$$(2.7) \qquad\qquad \bar{D} = c \circ \overline{\nabla}^S,$$

which acts on $\Gamma(\bar{S})$.

The bundles S and $\bar{S}$ are equipped with a natural hermitian structure. Let $d\,\mathrm{vol}_g$ denote the Riemannian volume form on M corresponding to the metric g. By $\mathcal{H}_e$ we denote the Hilbert space of square-integrable spinors with respect to the hermitian structure on S and the volume form $d\,\mathrm{vol}_g$, and $\mathcal{H}_p$ denotes the Hilbert space of square-integrable conjugate spinors.

Thus the manifold (M,g) gives rise to what A. Connes [1] calls *spectral triples*

$$(2.8) \qquad\qquad (\mathcal{A}, D, \mathcal{H}_e), \quad (\mathcal{A}, \bar{D}, \mathcal{H}_p).$$

[Note that D ($\bar{D}$) is selfadjoint on a dense domain in $\mathcal{H}_e$ ($\mathcal{H}_p$).] These spectral triples are familiar to anyone, who knows Pauli's non-relativistic quantum theory of the spinning electron and its twin, the positron:

(i) $\mathcal{A}$ is the algebra of electron (positron) position measurements;

(ii) $\mathcal{H}_e$ ($\mathcal{H}_p$) is the Hilbert space of pure state vectors of the electron (positron);

(iii) D ($\bar{D}$) is the square-root of the *Hamiltonian* generating the unitary time evolution of the states of an electron (positron) moving on (M,g) and coupled to an external magnetic field with field strength F_A (see eq. (2.5)). More precisely, the dynamics of an electron, with mass m and gyromagnetic factor g

---

[2]In our notation, we do not distinguish between a section, c, of $\mathrm{Cl}(M)$ and its representative as an endomorphism of $\Gamma(S)$ or of $\Gamma(\bar{S})$.

(measuring the strength of the magnetic moment of the electron) set equal to 2, is determined by the Hamiltonian

$$(2.9) \qquad H_A = \frac{\hbar^2}{2m} D_A^2,$$

where $\hbar$ is Planck's constant. In the presence of an electrostatic potential ϕ, the Hamiltonian is given by

$$(2.10) \qquad H_{\phi,A} := H_A + \phi, \quad \phi \in \mathcal{A}.$$

There is an isomorphism, C, between the spectral triples $(\mathcal{A}, D, \mathcal{H}_e)$ and $(\mathcal{A}, \overline{D}, \mathcal{H}_p)$, which the physicists call *charge conjugation*. The dynamics of a positron is determined by the Hamiltonian

$$(2.11) \qquad \overline{H}_{\phi,A} = \frac{\hbar^2}{2m} \overline{D}^2 - \phi.$$

The transition from $\mathcal{H}_e$ to $\mathcal{H}_p$, from D to $\overline{D}$, and from ϕ to $-\phi$ corresponds to reversing the sign of the electric charge, which converts an electron into a positron.

Obviously, the study of the spectral triples $(\mathcal{A}, D, \mathcal{H}_e)$ and $(\mathcal{A}, \overline{D}, \mathcal{H}_p)$ is thus a very natural enterprise from the point of view of the non-relativistic quantum theory of an electron or a positron, respectively. As Connes has shown, see [1], it is also a very natural starting point to explore the differential topology and Riemannian geometry of (M, g): The manifold M, as a topological Hausdorff space, is the spectrum of the abelian C^* algebra of continuous functions on M, which is the norm closure of $\mathcal{A}$; (Gel'fand's theorem). The differentiable structure on M can be reconstructed from the notion of $k = 1, 2, 3, \ldots$ times differentiable functions on M defined in terms of multiple commutators of $|D|$ with elements of $\mathcal{A}$. One continues by defining differential forms (in terms of certain sums of products of commutators of D with elements of $\mathcal{A}$), integration theory, $\ldots$. These constructions enable one to completely reconstruct the manifold (M, g) from the spectral triples (2.8); see Connes' book [1].

Next, we consider the isomorphism

$$(2.12) \qquad i : \overline{S} \otimes S (\otimes \mathbb{C}^2) \longrightarrow \Lambda^{\cdot} M,$$

where the factor $\mathbb{C}^2$ appears only if the dimension n of M is odd. The isomorphism i determines an isomorphism

$$(2.13) \qquad \mathcal{I} : \Gamma(\overline{S}) \otimes_{\mathcal{A}} \Gamma(S)(\otimes \mathbb{C}^2) \longrightarrow \Omega^{\cdot}(M)$$

from the space of tensor products of charge-conjugate spinors with spinors to differential forms. For even dimension n, we set

$$(2.14) \qquad \begin{aligned} \mathcal{I} \circ (\mathbf{1} \otimes c(\xi)) &:= \Gamma(\xi) \circ \mathcal{I}, \\ \mathcal{I} \circ (c(\xi) \otimes \gamma) &:= \overline{\Gamma}(\xi) \circ \mathcal{I}, \end{aligned}$$

for arbitrary $\xi \in \Omega^1(M)$, where γ is as in (2.3). For odd n,

$$(2.15) \qquad \begin{aligned} \mathcal{I} \circ (\mathbf{1} \otimes c(\xi) \otimes \tau_3) &=: \Gamma(\xi) \circ \mathcal{I}, \\ \mathcal{I} \circ (c(\xi) \otimes \mathbf{1} \otimes \tau_1) &=: \overline{\Gamma}(\xi) \circ \mathcal{I}, \end{aligned}$$

$$\text{where} \quad \tau_1 = \begin{pmatrix} 0 & 1 \\ 1 & 0 \end{pmatrix}, \qquad \tau_3 = \begin{pmatrix} 1 & 0 \\ 0 & -1 \end{pmatrix}.$$

The operators $\Gamma(\xi)$ and $\overline{\Gamma}(\xi)$ are anti-commuting sections of the Clifford bundle acting on $\Omega^{\cdot}(M)$. Defining

$$(2.16) \qquad \varepsilon(\xi) := \tfrac{1}{2}\left(\Gamma(\xi) - i\,\overline{\Gamma}(\xi)\right), \quad i(\xi) := -\tfrac{1}{2}\left(\Gamma(\xi) + i\,\overline{\Gamma}(\xi)\right),$$

one can verify that $\varepsilon(\xi)$ acts on $\Omega^{\cdot}(M)$ as *exterior multiplication* by ξ, while $i(\xi)$ acts on $\Omega^{\cdot}(M)$ as *interior multiplication* by the vector field, $X = G\xi$, dual to ξ.

Let $\mathcal{H}_{e-p}$ denote the Hilbert space of complex-valued, square-integrable differential forms. One can define two anti-commuting Pauli-Dirac operators, $\mathcal{D}$ and $\overline{\mathcal{D}}$, (selfadjoint on domains dense in $\mathcal{H}_{e-p}$) by setting

$$(2.17) \qquad\qquad \mathcal{D} = \Gamma \circ \nabla, \quad \overline{\mathcal{D}} = \overline{\Gamma} \circ \nabla.$$

They satisfy

$$(2.18) \qquad\qquad \{\mathcal{D}, \overline{\mathcal{D}}\} = 0, \quad \mathcal{D}^2 = \overline{\mathcal{D}}^2.$$

The relations (2.18) define what is called an $N = (1,1)$ *supersymmetry algebra* [2].

Thus, from the spectral triples of the electron and the positron (see (2.1), (2.6)–(2.8)), we can construct the $N = (1,1)$ *supersymmetric spectral data*

$$(2.19) \qquad\qquad \left(\mathcal{A}, \mathcal{D}, \overline{\mathcal{D}}, \mathcal{H}_{e-p}\right).$$

They can equivalently be described in terms of the operators d and d^* defined by

$$(2.20) \qquad\qquad d := \tfrac{1}{2}\left(\mathcal{D} - i\overline{\mathcal{D}}\right), \quad d^* := \tfrac{1}{2}\left(\mathcal{D} + i\overline{\mathcal{D}}\right),$$

acting on $\mathcal{H}_{e-p}$. It follows from (2.18), (2.20) that

$$(2.21) \qquad\qquad d^2 = (d^*)^2 = 0.$$

It turns out that, because the Levi-Civita connection ∇ has vanishing torsion, d is nothing but *exterior differentiation*.[3]

The passage from $(\mathcal{A}, D, \mathcal{H}_e)$ and $(\mathcal{A}, \overline{D}, \mathcal{H}_p)$ to $(\mathcal{A}, \mathcal{D}, \overline{\mathcal{D}}, \mathcal{H}_{e-p})$ has a natural *quantum-mechanical interpretation*: An electron and a positron can form bound states, (thanks to the Coulomb attraction, regularized at short distances). The *"groundstates"* of a bound electron-positron pair are called *positronium*. The Hilbert space of pure state vectors of positronium is precisely $\mathcal{H}_{e-p}$. The Hamiltonian describing the quantum-mechanical motion of positronium (i.e., the motion of the center of mass of the bound electron-positron pair) is given by

$$(2.22) \qquad\qquad H = \frac{\hbar^2}{2\mu}\,\mathcal{D}^2 = \frac{\hbar^2}{2\mu}\,\overline{\mathcal{D}}^2 = \frac{\hbar^2}{2\mu}\,(dd^* + d^*d),$$

where $\mu = 2m$ is the mass of positronium.

Note that $\mathcal{D}, \overline{\mathcal{D}}$ and H are *independent* of the U(1)-connection (vector potential) A, which, physically, corresponds to the circumstance that the electric charge of positronium is zero. The data $(\mathcal{A}, \mathcal{D}, \overline{\mathcal{D}}, \mathcal{H}_{e-p})$ are thus meaningful even if M does *not* admit a spin structure.

As verified in [2], de Rham-Hodge theory and the Riemannian geometry of (M, g) can be reconstructed completely from the $(1,1)$-supersymmetric spectral data $(\mathcal{A}, \mathcal{D}, \overline{\mathcal{D}}, \mathcal{H}_{e-p})$, supplemented by the operators T and $*$, where T counts the *degree* of a differential form, and $*$ is the unitary *Hodge involution*. In odd dimensions, $*$ anti-commutes with what the physicists call parity; in even dimensions, $*$ commutes with parity and is given by $* = \mathcal{I} \circ (\mathbf{1} \otimes \gamma) \circ \mathcal{I}^{-1}$, where γ is defined in (2.3). The quantum theory of positronium really completely encodes the de

---

[3]More generally, d is exterior differentiation "twisted" by a closed, odd differential form, [2].

Rham-Hodge theory and Riemannian geometry of (M,g). This theme is developed systematically in [2,3]. The point of view developed in [2,3] is, of course, inspired by [4] and [1].

It is interesting and important to explore how additional geometrical structure on M manifests itself in the quantum theory of an electron, a positron or of positronium. Let us imagine, for example, that M is a *symplectic* manifold, i.e., M is equipped with a globally defined, non-degenerate, closed 2-form ω. Let Ω denote the anti-symmetric bi-vector field obtained by inversion of ω. We define operators

$$(2.23) \qquad L_3 := T - \frac{n}{2}, \quad L_+ := \tfrac{1}{2}\varepsilon(\omega), \quad L_- := \tfrac{1}{2}i(\Omega),$$

where ε denotes exterior and i interior multiplication. Then

$$(2.24) \qquad [L_3, L_\pm] = \pm 2L_\pm, \quad [L_+, L_-] = L_3.$$

Thus L_3, L_+ and L_- span the Lie algebra sl_2. Because ω is closed, we have that $[L_+, d] = 0$. Defining a differential $\tilde{d}^*$ of degree -1 by

$$(2.25) \qquad \tilde{d}^* := [L_-, d],$$

we verify that

$$(2.26) \qquad (\tilde{d}^*)^2 = 0, \quad \{\tilde{d}^*, d\} = 0, \quad [L_-, \tilde{d}^*] = 0.$$

Thus $(d, \tilde{d}^*)$ transforms as a doublet under the adjoint action of sl_2.

It is known that every symplectic manifold M can be equipped with an almost complex structure J such that the tensor

$$(2.27) \qquad g(X,Y) := -\omega(JX, Y),$$

for arbitrary vector fields X and Y, defines a Riemannian metric on M. If $\Omega^\cdot(M)$ is completed to a Hilbert space $\mathcal{H}_{e-p}$ with respect to the norm determined by the metric g of (2.27) then L_3 is selfadjoint on $\mathcal{H}_{e-p}$ and

$$(2.28) \qquad (L_\pm)^* = L_\mp.$$

The operators L_3, L_+, L_- commute with the representation of $\mathcal{A}$ on $\mathcal{H}_{e-p}$. We can now define a second sl_2 doublet, $(\tilde{d}, -d^*)$, of differentials with the same properties as $(d, \tilde{d}^*)$. We are *not* claiming that

$$(2.29) \qquad \{d, \tilde{d}\} = 0;$$

this equation does not hold for general symplectic manifolds.

We have shown that if (M,g) is symplectic, with g as in (2.27), then it gives rise to spectral data

$$(2.30) \qquad (\mathcal{A}, d, d^*, \{L_3, L_+, L_-\}, *, \mathcal{H}_{e-p}),$$

with the properties discussed above. There are four supersymmetry generators, $d, d^*, \tilde{d},$ and $\tilde{d}^*$, satisfying

$$(2.31) \qquad d^2 = (d^*)^2 = \tilde{d}^2 = (\tilde{d}^*)^2 = 0, \quad \{d, \tilde{d}^*\} = 0.$$

It is natural to ask what is special about (M,g) if (2.29) holds, i.e., if $\{d, \tilde{d}\} = 0$. It turns out that, in this case, the almost complex structure J appearing in (2.27) is actually a *complex structure*, and the manifold (M, g, J), with ω as in (2.27), is a *Kähler manifold*. Defining

$$(2.32) \qquad \partial = \tfrac{1}{2}(d - i\tilde{d}), \quad \bar{\partial} = \tfrac{1}{2}(d + i\tilde{d}),$$

one verifies that

$$(2.33) \qquad \partial^2 = \bar{\partial}^2 = 0, \quad \{\partial, \bar{\partial}^{\#}\} = 0, \quad \{\partial, \partial^*\} = \{\bar{\partial}, \bar{\partial}^*\},$$

where $A^{\#} = A$ or A^*. The spectral data $\left(\mathcal{A}, d, \tilde{d}, T, *, \mathcal{H}_{e-p}\right)$ of a Kähler manifold have a U(1)-symmetry, with generator J_0, besides the SU(2) symmetry generated by $\{L_3, L_+, L_-\}$:

$$(2.34) \qquad\qquad [J_0, d] = -i\tilde{d}, \quad [J_0, \tilde{d}] = id.$$

The generator J_0 is determined by the complex structure J. The U(1)-symmetry generated by J_0 commutes with the representation of $\mathcal{A}$ on $\mathcal{H}_{e-p}$. The equations (2.34) imply (2.29).

It may happen that, for certain manifolds (M, g), only the U(1)-symmetry with generator J_0 as in (2.34) is realized; but the SU(2)-symmetry characteristic of symplectic manifolds is *broken*. In this case, (2.29) holds, but $\{d, \tilde{d}^*\}$ does *not* vanish. One may then still introduce Dolbeault differentials, as in (2.32), but ∂ need not anti-commute with $\bar{\partial}^*$, and the two Kähler Laplacians, $\{\partial, \partial^*\}$ and $\{\bar{\partial}, \bar{\partial}^*\}$, may be different from each other. Manifolds (M, g) with these properties are *complex-hermitian* manifolds.

One can now go on and reformulate the geometries of hyper-Kähler and hyper-complex manifolds in terms of spectral data of the supersymmetric quantum theory of positronium, enriched by *unitary symmetries commuting with the representation of the algebra* $\mathcal{A}$, but transforming the differentials non-trivially. This program has been carried out in [2,3]. Moreover, it is outlined in these papers (and in the literature quoted there) how the approach to geometry described here is connected to the study of supersymmetric, non-linear σ-models and superconformal field theories (by dimensional reduction), and how to generalize it to *non-commutative* spaces.

Rather than studying the symmetries of a supersymmetry algebra (i.e., of the algebra of differentials $d, \tilde{d}, \ldots$) commuting with the algebra $\mathcal{A}$, as above, one may also study the symmetries of $\mathcal{A}$ commuting with some of the differentials: The *automorphism group $\mathrm{Diff}(M)$, of $\mathcal{A}$ is given by the *diffeomorphisms* of the manifold M. It can be represented on $\mathcal{H}_{e-p}$. This representation obviously commutes with exterior differentiation, i.e., with the differential d. The subgroup of $\mathrm{Diff}(M)$ commuting with d *and* with d^* consists precisely of diffeomorphisms of M that preserve the *metric* g on M, (i.e., of *Killing diffeomorphisms*). If (M, ω) is symplectic then the subgroup of $\mathrm{Diff}(M)$ commuting with d *and* with $\tilde{d}^*$ consists of *symplectomorphisms*, and if (M, J) is complex the subgroup of $\mathrm{Diff}(M)$ commuting with d and $\tilde{d}$ consists of (anti-)holomorphic diffeomorphisms, etc. ; see [2]. It is interesting to study the problems in how far the manifold M can be reconstructed from the supersymmetry algebra and from the group of those *automorphisms of $\mathcal{A}$ commuting with some family of differentials and how to generalize these notions to non-commutative spaces.

The punch line, in this section, is that the differential topology and geometry of classical manifolds can be completely encoded in (and reconstructed from) the *supersymmetric spectral data* provided by the *quantum theory of an electron, a positron, or of positronium* moving on such manifolds.

What do we gain by adopting this point of view? The answer is that we gain *generality* and *generalizations of geometry*. The quantum-theoretical approach to differential geometry enables us to study highly singular and discrete geometrical spaces and suggests an approach to the study of *non-commutative geometrical spaces*; (one then replaces $\mathcal{A} = C^\infty(M)$ by a non-commutative *algebra). Non-commutative geometry has been conceived and developed by Connes [1], (with the above purposes in mind).

3. Analysis of many-electron systems

After having indicated how differential topology and geometry are encoded into the supersymmetric quantum theory of electrons, positrons and positronium, we propose to describe some of the impact of the study of many-electron systems on *analysis*. The right speakers to discuss this matter would be C. Fefferman and E. H. Lieb, and excellent references to learn about it (and much more) are [5, 6] (and references given there).

3.1. Stability of matter. Electrons and positrons are fermions, i.e., they obey Pauli's exclusion principle; (positronium is a boson). The Hilbert space of pure state vectors of a system consisting of N non-relativistic electrons is thus given by

$$(3.1) \qquad \mathcal{H}^{(N)} := (\mathcal{H}_e)^{\wedge N} = \mathcal{H}_e \wedge \cdots \wedge \mathcal{H}_e,$$

where $\wedge$ denotes an anti-symmetric tensor product, and there are N factors on the right-hand side of (3.1). The Hamiltonian, $H^{(N)}_{\phi,A,v}$, generating the time evolution of an N-electron system is given by

$$(3.2) \qquad H^{(N)}_{\phi,A,v} := \sum_{i=1}^{N}\Big(\mathbf{1} \otimes \cdots \otimes \underset{\uparrow j}{H_{\phi,A}} \otimes \cdots \otimes \mathbf{1}\Big) + \sum_{i<j} v_{ij},$$

where $H_{\phi,A}$ is the Hamiltonian introduced in (2.10), and $v_{ij} = v(x_i, x_j)$ is a two-body potential; x_i denotes the position of the ith electron, and v is a function on $M \times M$ which is smooth except, possibly, on the diagonal ($x_i = x_j$).

We are interested in studying properties of $H^{(N)}_{\phi,A,v}$ (selfadjointness, location and nature of spectrum, etc.). The simplest, *non-trivial* problem is to consider a single electron, as in Section 2. We imagine that ϕ is the electrostatic potential created by some pointlike nucleus of charge Z located at a point $p \in M$; (we use units such that the elementary electric charge $e = 1$; Z is the atomic number). Then

$$(3.3) \qquad \phi(x) = -ZG(x, p),$$

where $G(x, p)$ is the Green function of the scalar Laplace-Beltrami operator; (x is the position of the electron). Due to the interaction of the magnetic moment of the electron with the magnetic field F_A (see eq. (2.5)), i.e., due to the Zeeman energy,

$$(3.4) \qquad \inf_A \,(\inf \operatorname{spec} H_{\phi,A}) = -\infty,$$

in dimension $n \geq 2$; see [7] and references given there. Thus, an electron in the presence of a static nucleus and under the influence of suitable, static external magnetic fields can be in a state of arbitrarily negative energy. This result is less discouraging than it may seem: In (3.4), we have neglected the energy of the magnetic field, which is given by

$$(3.5) \qquad \mathcal{E}_f(A) := \Gamma \int_M F_A \wedge F_A^*,$$

where Γ is proportional to α^{-2}, and α is the feinstructure constant; (in three dimensions, $\alpha \simeq 1/137$ is dimensionless). We really ought to consider the energy functional

$$(3.6) \qquad \mathcal{E}(\psi, A) := \langle \psi, H_{\phi, A} \psi \rangle + \mathcal{E}_f(A),$$

where ψ ranges over a dense subspace of $\mathcal{H}_e$ and has unit norm, $\|\psi\|_{\mathcal{H}_e} = 1$. We are interested in estimating the infimum of $\mathcal{E}(\psi, A)$ over all $\psi \in \mathcal{H}_e$, with $\|\psi\|_{\mathcal{H}_e} = 1$, and over all A. Simple dimensional analysis suggests that if the dimension, n, of M is $n \geq 4$ then $\inf \mathcal{E}(\psi, A) = -\infty$; and for $n = 2$, $\inf \mathcal{E}(\psi, A) \geq C_\Gamma > -\infty$, for some finite constant C_Γ. An interesting problem is encountered when $n = 3$, which is the dimension of physical space. Then one can show that there is a critical value, $\Gamma_c = \Gamma_c(Z)$ (increasing in Z), such that

(i) for $\Gamma > \Gamma_c$, $\displaystyle\inf_{\psi, A} \mathcal{E}(\psi, A) > -\infty$;

(ii) for $\Gamma < \Gamma_c$, $\displaystyle\inf_{\psi, A} \mathcal{E}(\psi, A) = -\infty$;

see [7, 8, 9]. This result (in particular, part (ii)) is a consequence of the existence of *zero-modes* for the Pauli-Dirac operator $D \equiv D_A$ introduced in (2.6): Let ψ_A be a solution of the equation

$$(3.7) \qquad D_A \psi = 0,$$

with $\|\psi_A\|_{\mathcal{H}_e} = 1$.

Then, by (2.10), (3.7) and (3.3),

$$(3.8) \qquad \mathcal{E}(\psi_A, A) = -Z \langle \psi_A, G(\,\cdot\,, p)\psi_A \rangle + \mathcal{E}_f(A).$$

Let l denote length. It is easy to verify that the first term on the right-hand side of (3.8) scales like l^{2-n}, while the second term scales like l^{n-4}, for $l \to 0$. In the "critical" dimension $n = 3$, the two terms both scale like l^{-1}. One then finds that

$$(3.9) \qquad Z\Gamma_c^{-1} = \inf_{\mathcal{C}} \left(\mathcal{E}_f(A) \,/\, \langle \psi, G(\,\cdot\,, p)\psi \rangle \right) > 0,$$

where $\mathcal{C}$ consists of all configurations (A, ψ) with $\mathcal{E}_f(A) < \infty$, $D_A \psi = 0$, $\|\psi\|_{\mathcal{H}_e} = 1$; see [7, 8]. The key problem is thus to understand whether eq. (3.7) has normalizable solutions, $\psi = \psi_A$, for vector potentials A with $\mathcal{E}_f(A) < \infty$. In order to tackle this problem, M. Loss and H.-T. Yau [9] have constructed solutions to the equations

$$D_A \psi = 0, \quad c(F_A) = g(\psi \cdot \psi^*)_0,$$

where g is a numerical constant, c is the map from differential forms to sections of the Clifford bundle (as in Section 2), and K_0 denotes the traceless part of a matrix K.

These equations are the three-dimensional analogues of the celebrated Seiberg-Witten equations [10] which have revolutionized four-dimensional topology.

In attempting to understand the properties of non-relativistic matter, we must, of course, study systems consisting of many electrons. To make contact with laboratory quantum physics, we choose physical space M to be three-dimensional

Euclidian space, $\mathbb{E}^3$. In (3.2), we choose $\phi = \phi^{(K)}$ to be the electrostatic potential created by K pointlike, static nuclei of atomic numbers $Z_1, \ldots, Z_K$ located at points $p_1, \ldots, p_K$ in $\mathbb{E}^3$. We assume that $Z_j \leq Z$, $j = 1, \ldots, K$, for some (arbitrary, but) fixed constant Z independent of K and of the positions $p_1, \ldots, p_K$. Let

$$E^{(K)}(\underline{p}) = \sum_{i<j} \frac{Z_i \, Z_j}{4\pi |p_i - p_j|}$$

denote the electrostatic energy of the configuration of nuclei. We set

$$v_{ij} = v(x_i, x_j) = \frac{1}{4\pi |x_i - x_j|}.$$

We define an energy functional $\mathcal{E}^{(N,K)}$ by

$$(3.10) \qquad \mathcal{E}^{(N,K)}(\psi, A) := \langle \psi, H^{(N)}_{\phi^{(K)}, A, v} \psi \rangle + E^{(K)}(\underline{p}) + \mathcal{E}_f(A),$$

where $\psi \in \mathcal{H}^{(N)}$ (see (3.1)), with $\|\psi\|_{\mathcal{H}^{(N)}} = 1$, and $\mathcal{E}_f(A)$ is as in (3.5). A key physical problem is to show that

$$(3.11) \qquad \inf_{(A, \psi) \in \mathcal{C}^{(N)}} \mathcal{E}^{(N,K)}(\psi, A) \geq -C(N + K),$$

for some finite constant $C \geq 0$, provided Z is sufficiently small and the (dimensionless) constant Γ in (3.5) is large enough. In (3.11), $\mathcal{C}^{(N)}$ consists of all configurations (A, ψ) such that $\mathcal{E}_f(A) < \infty$ and $\psi \in \mathcal{H}^{(N)}$, $\|\psi\|_{\mathcal{H}^{(N)}} = 1$. Inequality (3.11) says that the energy per particle is bounded below, uniformly in the number, N, of electrons and the number, K, of nuclei, ("*stability of matter*"). There are many variants of this problem; see [5]. The first proof of a result similar to (3.11) was found by Dyson and Lenard. Later, Lieb and Thirring discovered a beautiful, new proof of the Dyson-Lenard theorem based on a new type of Sobolev inequality involving the Pauli principle; see [5]. Inequality (3.11) was conjectured in [7] and was recently proven by Fefferman [11] and by Lieb, Loss and Solovej [12].

So far, the magnetic field described by the vector potential A has been treated as a classical, static external field. In many situations of condensed matter physics, this appears to be a good approximation. But one should remember that quantum theory was originally discovered by studying the electromagnetic field in a cavity (black body radiation) and finding that it has *quantum-mechanical* properties. Thus, we should really treat the electromagnetic field quantum-mechanically. The Hilbert space of pure state vectors of the quantized electromagnetic field (in the so-called Coulomb gauge) is the Fock space, $\mathcal{F}$. [It is the symmetric tensor algebra over $L^2(\mathbb{R}^3, d^3k) \otimes \mathbb{C}^2$, where $\mathbb{R}^3$ is momentum space of a single photon, and $\mathbb{C}^2$ is the state space for the helicities (polarizations) of a photon.] It is convenient to introduce creation- and annihilation operators $a_\lambda^*(k)$, $a_\lambda(k)$, $k \in \mathbb{R}^3$, $\lambda = \pm$, satisfying the Heisenberg algebra

$$[a_\lambda^\#(k), a_{\lambda'}^\#(k')] = 0 \quad (a^\# = a \text{ or } a^*),$$
$$(3.12) \qquad [a_\lambda(k), a_{\lambda'}^*(k')] = \delta_{\lambda\lambda'} \, \delta(k - k').$$

For $\varphi_\lambda(k) \in L^2(\mathbb{R}^3, d^3k) \otimes \mathbb{C}^2$,

$$a(\varphi) := \sum_{\lambda=\pm} \int d^3k \, \varphi_\lambda(k) \, a_\lambda(k)$$

and its adjoint, $a(\varphi)^* = a^*(\overline{\varphi})$, are densely defined, unbounded operators on $\mathcal{F}$. Fock space contains a distinguished vector, the *vacuum* (vector), $|0\rangle$ (unique, up to a complex phase), with the property that

$$(3.13) \qquad\qquad a(\varphi)|0\rangle = 0,$$

for all $\varphi \in L^2(\mathbb{R}^3, d^3k) \otimes \mathbb{C}^2$. The quantized vector potential (in the Coulomb gauge) is given by

$$(3.14) \quad A(x) := \frac{1}{\sqrt{(2\pi)^3}} \sum_{\lambda=\pm} \int \frac{d^3k}{\sqrt{2|k|}} \left\{ a_\lambda^*(k)\, \varepsilon_\lambda(k)\, e^{-ikx} + a_\lambda(k)\, \overline{\varepsilon_\lambda(k)}\, e^{ikx} \right\},$$

where $\varepsilon_+(k)$ and $\varepsilon_-(k)$ are unit (polarization) vectors in $\mathbb{R}^3 \otimes \mathbb{C}$ with the property that $\left(\frac{k}{|k|}, \varepsilon_+(k), \varepsilon_-(k) \right)$ form an orthonormal basis of $\mathbb{R}^3 \otimes \mathbb{C}$. The field $A(x)$ is an unbounded, operator-valued distribution on $\mathcal{F}$. In order to base an analysis of the quantum theory of electrons interacting with the quantized radiation field on a mathematically firm starting point, we must regularize $A(x)$. We define a vector potential cutoff in the ultraviolet by

$$(3.15)$$
$$A^{(\Lambda)}(x) := \frac{1}{\sqrt{(2\pi)^3}} \sum_{\lambda=\pm} \int \frac{d^3k}{\sqrt{2|k|}} \, \Lambda(k)\left\{ a_\lambda^*(k)\varepsilon_\lambda(k)\, e^{-ik\cdot x} + a_\lambda(k)\, \overline{\varepsilon_\lambda(k)}\, e^{ik\cdot x} \right\}$$

where $\Lambda(k)$ is the characteristic function of the ball $\{ k \mid |k| \leq \Lambda \} \subset \mathbb{R}^3$.

The time evolution of the free electromagnetic field is generated by the Hamiltonian

$$(3.16) \qquad\qquad H_f = \sum_{\lambda=\pm} \Gamma \int d^3k \, a_\lambda^*(k)|k|\, a_\lambda(k)$$

which is a positive operator selfadjoint on a dense domain in $\mathcal{F}$. By (3.13), $H_f|0\rangle = 0$. The spectrum of H_f consists of a simple eigenvalue, 0, corresponding to the eigenvector $|0\rangle$ and an absolutely continuous part of infinite multiplicity covering the positive half-axis.

The Hilbert space of a system consisting of N (non-relativistic) electrons interacting with the quantized electromagnetic field is given by

$$(3.17) \qquad\qquad \mathcal{H} := \mathcal{H}^{(N)} \otimes \mathcal{F},$$

where $\mathcal{H}^{(N)}$ has been defined in (3.1). The Hamiltonian of the system, in the presence of K pointlike, static nuclei, is defined by

$$(3.18) \qquad\quad H^{(N,K;\Lambda)} := H^{(N)}_{\phi^{(K)}, A^{(\Lambda)}, v} + E^{(K)}(\underline{p}) + H_f,$$

where $H^{(N)}_{\phi^{(K)}, A^{(\Lambda)}, v}$ has been defined in (3.2), with $\phi^{(K)}$ as above, $A^{(\Lambda)}$ as in (3.15), v as in (3.9), and where H_f is given by (3.16). It is not difficult to show that, for $N < \infty$, $\Lambda < \infty$, $H^{(N,K;\Lambda)}$ is a well defined operator which is bounded from below and selfadjoint on a dense domain in $\mathcal{H}$. It is highly non-trivial to show that, for $Z < \infty$ and $\Gamma > 0$,

$$(3.19) \qquad\qquad H^{(N,K;\Lambda)} \geq -C_\Lambda\, (N + K),$$

for some finite constant C_Λ. In contrast to (3.11), we do *not* have to assume, here, that Z is small and Γ is large enough. This is due to the fact that, for $\Lambda < \infty$,

$A^{(\Lambda)}(x)$ is smooth in x. The prize to pay is that the constant C_Λ blows up, as $\Lambda \nearrow \infty$. A proof of (3.19) can be found in [13]; see also [14].

The quantum theory with the Hilbert space $\mathcal{H}$ and the Hamiltonian $H^{(N,K;\Lambda)}$ is believed to describe much of chemistry, spectroscopy and quantum optics, and condensed matter physics. Experimental data are well reproduced by the theory; but there are corrections due to the motion of the nuclei and to a variety of relativistic effects. However, experimental data do not appear to exhibit any significant dependence on the regularization parameter (ultra-violet cutoff) Λ. This suggests that one ought to be able to construct a *renormalized* Hamiltonian, $H_{\text{ren}}^{(N,K)}$, as a limit of the Hamiltonians $H^{(N,K;\Lambda)}$, as $\Lambda \to \infty$. This problem already worried the founding fathers of quantum mechanics and the participants of the Shelter Island conference. From the point of view of rigorous analysis, it remains open! But there has been significant progress, recently, towards understanding how its solution should look like: Recall that the one-electron Hamiltonian $H_{\phi,A}$ is given by

$$H_{\phi,A} = \frac{\hbar^2}{2m} D_A^2 + \phi;$$

see eqs. (2.10) and (2.11). We must choose m to depend on the ultraviolet cutoff Λ: $m = m(\Lambda)$. There is a mathematical technique, called *renormalization group*, that was pioneered by Wilson, see [15], that can be used to study the correct Λ-dependence of $m(\Lambda)$. It suggests to choose

$$(3.20) \qquad m(\Lambda) = m_0 (m_0 / \Lambda)^{c_1 \alpha + O(\alpha^2)}$$

where m_0 is roughly given by the experimentally measured value of the mass of the electron, $\alpha \simeq 1/137$ is the feinstructure constant, and $c_1 > 0$ is a computable numerical constant, [16]. We define

$$(3.21) \qquad H_{\phi,A^{(\Lambda)}}^{(\Lambda)} = \frac{\hbar^2}{2m(\Lambda)} D_{A^{(\Lambda)}}^2 + \phi - \mu_0(\Lambda),$$

where $m(\Lambda)$ is as in (3.20), and $\mu_0(\Lambda) \sim O(\alpha\Lambda^2)$ is some energy scale depending on m_0, α and Λ. In eq. (3.2), we replace $H_{\phi,A}$ by $H_{\phi,A^{(\Lambda)}}^{(\Lambda)}$ to define an operator $H_{\phi,A^{(\Lambda)},v}^{(N;\Lambda)}$. In (3.16), we choose $\Gamma \propto \alpha^{-2}$, and, in (3.18), we replace $H_{\phi(K),A^{(\Lambda)},v}^{(N)}$ by $H_{\phi(K),A^{(\Lambda)},v}^{(N,\Lambda)}$. This yields a family of operators, $H_{\text{ren}}^{(N,K;\Lambda)}$, depending on the ultraviolet cutoff Λ. The conjecture is that if α and Z are *sufficiently small*, and for $m(\Lambda)$ as in (3.20) and a suitable choice of $\mu_0(\Lambda)$,

$$(3.22) \qquad H_{\text{ren}}^{(N,K;\Lambda)} \geq -C(N+K),$$

for some finite constant C *independent* of Λ. Although various important elements for a proof of (3.22) are visible, we are presumably miles away from a rigorous proof of this inequality.

So far, we have described purely foundational results on the quantum theory of systems consisting of many electrons interacting with the quantized radiation field. To make contact with experimental data, we have to work out more detailed, quantitative consequences of the theory. There are many natural, important questions that one may ask and that remain open. I propose to describe some of them.

3.2. Stability of atoms and molecules. Suppose we turn off the interactions between electrons and the electromagnetic field (i.e., we set $\Lambda = 0$). Then we understand much about the stability and constitution of atoms and ions with a single nucleus; (the quantum-mechanical motion of the nucleus is taken into account, in the theory). See [5] and refs. given there. We also understand some of the properties of the simplest gas consisting of hydrogen (electrons and protons) [17, 5]. But, for purposes of quantum chemistry, we should like to understand the stability and properties of *molecules* and the nature of *chemical binding*. Assuming that we incorporate the motion of nuclei and that we insist on mathematical rigor, our understanding of molecules is really very rudimentary: Only recently, the stability and some properties of molecules consisting of two nuclei and one or two electrons have been established; see [18]. There are no rigorous results beyond these, unless one is willing to make the Born-Oppenheimer approximation; (ratio between electron mass and mass of nuclei $\to 0$).

3.3. Open quantum systems and quantum friction. Quantum theory was discovered by studying the properties of black-body radiation and *atomic spectroscopy*. How much do we understand about spectroscopy, i.e., about the radiation theory of atoms (and molecules), assuming we want to go beyond perturbation theory and insist on mathematical rigor? There has been very little work on this question, until recently. In [19], a rigorous approach to atomic spectroscopy has been initiated under the simplifying assumptions that the nucleus of the atom is static, an ultra-violet cutoff $\Lambda < \infty$ is imposed (and that interactions between electrons very *far* from the nucleus and the radiation field can be neglected). Although these assumptions are physically reasonable, one should ultimately get rid of them! The Hamiltonian of the system is given by (3.18), with $K = 1$ ($\phi^{(1)} \equiv \phi$), and $A^{(\Lambda)}(x)$ is replaced by

$$(3.23) \qquad\qquad a(x) = A^{(\Lambda)}(x)\, g(x),$$

where $g(x) > 0$ is smooth, $g(0) = 1$, and $g(x) \to 0$ "rapidly", as $|x| \to \infty$.[4] This Hamiltonian can be written as

$$(3.24) \qquad\qquad H_{\text{tot}} = H^{(N)}_{\phi, a, v} + H_f := H_0 + I,$$

where

$$(3.25) \qquad\qquad H_0 = H^{(N)}_{\phi,\, a=0, v} + H_f.$$

The Hamiltonian $H^{(N)}_{\phi, a=0, v}\big|_{\mathcal{H}^{(N)}}$ describes an N-electron atom or positive ion; its spectrum consists of discrete eigenvalues $e_0 < e_1 < e_2 < \cdots$ of finite multiplicity and continuous spectrum contained in $[\Sigma, \infty)$, where Σ is the ionization threshold. The boundstate energies e_j, $j = 0$, (groundstate), $1, 2, \ldots$ (excited states) are eigenvalues of H_0 with the same multiplicity as before; but a branch of continuous spectrum of infinite multiplicity is attached to every e_j. Thus, for $j > 0$, e_j is an *embedded eigenvalue* of H_0. The term I on the right-hand side of (3.24) describes the interactions between the electrons and the radiation field; it is $O(\alpha^{1/2})$, where α is the feinstructure constant. One would like to treat I perturbatively. The physical effect caused by adding I is to convert the eigenvalues $e_j, j > 0$, into complex resonance energies. The corresponding unperturbed eigenstates become *unstable states of finite life time*. In [19], V. Bach, I. M. Sigal and I have attempted

---

[4]Many of the results reported below still hold when $g(x) \equiv 1$.

to rigorously establish this picture (under the simplifying assumptions described above). Furthermore, we have shown that inf spec H_{tot} remains an eigenvalue of H_{tot} of finite multiplicity; (*existence of groundstates*). These results are proven for sufficiently small values of α.

Among analytical techniques used to prove these results are the following ones.

(a) Complex dilatations: Complex dilatations of physical space and photon momentum space are represented on the Hilbert space $\mathcal{H} = \mathcal{H}^{(N)} \otimes \mathcal{F}$ by densely defined, unbounded operators $D(\theta)$, where θ denotes the dilatation parameter. As long as θ is real, $D(\theta)$ is unitary. One considers the family of operators

$$(3.26) \qquad H_{\text{tot}}(\theta) := D(\theta)^{-1} \, H_{\text{tot}} \, D(\theta)$$

and verifies that it is an analytic family of type A (see [20]), for θ in a sufficiently narrow strip around the real axis. For Im $\theta \neq 0$, the spectrum of the unperturbed operator $H_0(\theta)$ now consists of isolated branches, $\mathcal{B}_j^{(\theta)}$, of continuous spectrum of the form

$$(3.27) \qquad \mathcal{B}_j^{(\theta)} = \left\{ z \;\middle|\; \arg\,(z - e_j) = \text{Im } \theta \right\}$$

and a continuum $\mathcal{B}_\Sigma$ described by

$$\mathcal{B}_\Sigma^{(\theta)} = \left\{ z \;\middle|\; 0 \leq \frac{\arg\,(z - \Sigma)}{\text{Im } \theta} \leq 2 \right\}.$$

The perturbation $I(\theta) = D(\theta)^{-1} \, I \, D(\theta)$ is small relative to $H_0(\theta)$, in the sense of Kato [20], for sufficiently small α.

The problem of resonances in atomic spectroscopy is to understand how the half-lines $\mathcal{B}_j^{(\theta)}$, $j = 0, 1, 2, \ldots$, are deformed when the perturbation $I(\theta)$ is added. For $I(\theta) = 0$, e_j is a real eigenvalue of the operator $H_0(\theta)$ attached to a branch $\mathcal{B}_j^{(\theta)}$ of continuous spectrum. It is therefore *not* possible to apply simple, analytic perturbation theory to study the behavior of e_j and $\mathcal{B}_j^{(\theta)}$ when the perturbation $I(\theta)$ is added; more sophisticated techniques are needed to tackle problems of this kind. In [19], V. Bach, I. M. Sigal and I have developed a new variant of the renormalization group method to study the fate of e_j and of $\mathcal{B}_j^{(\theta)}$ when $I(\theta)$ is turned on, for $\theta = -i\varphi$, $\varphi > 0$. It is based on:

(b) The Feshbach map: Let $\mathcal{H}$ be a separable Hilbert space, H a densely defined, closed operator on $\mathcal{H}$, and let P be a bounded projection, and $\overline{P} := \mathbf{1} - P$, with the following properties: The domain of H contains the range of P; we may therefore define $H_P := PHP$ and $H_{\overline{P}} := \overline{P}H\overline{P}$. We assume that for z in the resolvent set, $\rho(H_{\overline{P}})$, of $H_{\overline{P}}$

$$(3.28) \qquad \| \overline{P}(H_{\overline{P}} - z)^{-1} \, \overline{P}HP \| < \infty, \quad \| PH\overline{P}\,(H_{\overline{P}} - z)^{-1}\overline{P} \| < \infty.$$

For H and P as above, and $z \in \rho(H_{\overline{P}})$, we define the Feshbach map

$$(3.29) \qquad f_{z,P} : H \;\mapsto\; f_{z,P}(H) := \left\{ PHP - PH\overline{P}(H_{\overline{P}} - z)^{-1} \, \overline{P}HP \right\} \Big|_{\text{Ran } P}.$$

The properties of this map are summarized in the following theorem [19].

Theorem. *Under the hypothesis* (3.28), *and for* $z \in \rho(H_{\overline{P}})$, *the Feshbach map* (3.29) *is well defined, and*

(i) $z \in \rho(H) \Longleftrightarrow z \in \rho\,(f_{z,P}(H))$,

(ii) $z \in \sigma_{pp}(H) \Longleftrightarrow z \in \sigma_{pp}\left(f_{z,P}(H)\right)$,

where $\sigma_{pp}(\,\cdot\,)$ denotes the pure point spectrum (eigenvalues); and

$$\dim \ker (H - z) = \dim \ker \left(f_{z,P}(H) - z\right).$$

(iii) If P_1 and P_2 commute with one another then

$$f_{z,P_1} \circ f_{z,P_2} = f_{z,P_1 P_2}.$$

(c) **Application:** Let $P_{\mathrm{at}}^{(j)}(\theta)$ denote the projection onto the eigenspace of the dilated atomic Hamiltonian $H_{\phi,a=0,v}^{(N)}(\theta)\big|_{\mathcal{H}^{(N)}}$ corresponding to the eigenvalue e_j, and let $P_{H_f \leq \rho_0}$ denote the spectral projection of H_f onto the interval $[0,\rho_0]$, for some constant $\rho_0 < \min\left(e_j - e_{j-1},\ e_{j+1} - e_j,\ \Sigma - e_j\right)$ to be chosen later. Let $P^{(0)} = P_{\mathrm{at}}^{(j)}(\theta) \otimes P_{H_f \leq \rho_0}$. We fix $\theta = -i\varphi$, $\varphi > 0$ (small enough), and we assume that $z \in \rho\left(H_{\mathrm{tot}}(\theta)_{\overline{P}^{(0)}}\right)$. Then the Feshbach map

$$(3.30) \qquad H_{\mathrm{tot}}(\theta) \mapsto f_{z,P^{(0)}}\left(H_{\mathrm{tot}}(\theta)\right) := H^{(0)}(z)$$

defines a bounded operator $H^{(0)}(z)$ on the Hilbert space

$$(3.31) \qquad \mathcal{H}^{(0)} := \mathrm{Ran}\, P^{(0)} \simeq \mathbb{C}^{n_j} \otimes \mathcal{F}^{(0)},$$

where n_j is the multiplicity of the eigenvalue e_j of H_0, and $\mathcal{F}^{(0)} := P_{H_f \leq \rho_0} \mathcal{F}$. For small enough values of the feinstructure constant α and for small enough ρ_0, the complex disk

$$\mathcal{S}_0 := \left\{ z \mid |z - e_j| < \tfrac{1}{2}\rho_0 \right\}$$

is contained in the resolvent set $\rho\left(H_{\mathrm{tot}}(\theta)_{\overline{P}^{(0)}}\right)$ of $H_{\mathrm{tot}}(\theta)_{\overline{P}^{(0)}}$, and we can explore the part of the spectrum of $H_{\mathrm{tot}}(\theta)$ intersecting $\mathcal{S}_0$ by studying the spectrum of $H^{(0)}(z)$ inside $\mathcal{S}_0$, $(z \in \mathcal{S}_0)$. It is shown in [19] that if *Fermi's Golden Rule* (to second order in the perturbation I) predicts that the eigenvalue e_j turns into n_j complex resonance energies with *strictly negative imaginary parts* then, for sufficiently small α, there is a choice of the constant ρ_0 such that $\mathcal{S}_0 \cap \mathbb{R}$ does *not contain any spectrum* of the operator $H^{(0)}(z)$. Of course, this is equivalent to claiming that $\mathcal{S}_0 \cap \mathbb{R}$ does not contain any spectrum of $H_{\mathrm{tot}}(\theta)$; but the operator $H^{(0)}(z)$ turns out to be easier to analyze than $H_{\mathrm{tot}}(\theta)$, because $H^{(0)}(z)$ can be shown to be a small perturbation of the operator

$$(3.32) \qquad \varepsilon^{(0)}(z) \otimes \mathbf{1} + \mathbf{1} \otimes e^{-i\varphi}\, H_f$$

acting on $\mathcal{H}^{(0)} \simeq \mathbb{C}^{n_j} \otimes \mathcal{F}_0$, where $\varepsilon^{(0)}(z)$ is a diagonal $n_j \times n_j$ matrix with a *strictly negative* imaginary part, for $z \in \mathcal{S}_0$.

If one desires to know more precisely where the spectrum of $H^{(0)}(z)$ (and thus of $H_{\mathrm{tot}}(\theta)$) inside $\mathcal{S}_0$ is located one continues by applying a Feshbach map to $H^{(0)}(z)$. One defines $\mathcal{S}_1 \subset \mathcal{S}_0$ to be the disk of radius $\frac{1}{2}M^{-1}\rho_0$ centered at the bary center of a group of k eigenvalues of $\varepsilon^{(0)}(z)$, $1 \leq k \leq n_j$, separated from each other by a distance $\ll M^{-1}\rho_0$ and corresponding to a k-dimensional projection $P_k^{(j)}$ on $\mathbb{C}^{n_j}$; here $M \gg 1$ is a constant that will be kept fixed. We define

$$P^{(1)} := P_k^{(j)} \otimes P_{H_f \leq M^{-1}\rho_0}.$$

One now attempts to verify that $\mathcal{S}_1$ is in the resolvent set of $H^{(0)}(z)_{\overline{P}^{(1)}}$. This enables one to define the Feshbach Hamiltonian $f_{z,P^{(1)}}\left(H^{(0)}(z)\right)$. Let $D_f(\ln M)$

denote the unitary dilatation operator on $\mathcal{F}$ mapping the range of $P_{H_f \leq M^{-1}\rho_0}$ onto $\mathcal{F}^{(0)}$. For $z \in \mathcal{S}_1$, we define

$$H^{(1)}(z) := M \, D_f(\ln M) \, f_{z,P^{(1)}}\left(H^{(0)}(z)\right) D_f(\ln M)^{-1}.$$

This is an operator on the Hilbert space

$$\mathrm{Ran} \, P_k^{(j)} \otimes \mathcal{F}^{(0)}.$$

We are interested in locating the spectrum of $H^{(1)}(z)$ inside the disk $M \, \mathcal{S}_1$ (and thus of $H_{\mathrm{tot}}(\theta)$ inside the disk $\mathcal{S}_1$!). This is accomplished by showing that $H^{(1)}(z)$ is a small perturbation of the operator

$$\varepsilon^{(1)}(z) \otimes \mathbf{1} + \mathbf{1} \otimes e^{-i\varphi} \, H_f$$

on the space $\mathrm{Ran} \, P_k^{(j)} \otimes \mathcal{F}^{(0)}$.

The construction described here is then iterated. The goal is to show that there are complex numbers $\varepsilon_1, \ldots, \varepsilon_{n_j}$ of strictly negative imaginary part such that, on spaces $V^{(l)} \otimes \mathcal{F}^{(0)}$, and for certain sequences $\{z_n\}_{n=0}^{\infty}$ converging to complex numbers ε_l (with $\mathrm{Im}\,\varepsilon_l < 0$), $l = 1, \ldots, n_j$,

$$H^{(n)}(z_n) - \left(M^n \varepsilon_l \otimes \mathbf{1} + \mathbf{1} \otimes e^{-i\varphi} \, H_f\right) \to 0,$$

as $n \to \infty$, where $V^{(l)}$ is a one-dimensional subspace of $\mathbb{C}^{n_j}$, for $l = 1, \ldots, n_j$, and $\oplus V^{(l)} = \mathbb{C}^{n_j}$. The numbers $\varepsilon_1, \ldots, \varepsilon_{n_j}$ are complex resonance energies; to leading order, their real parts are given by Bethe's formulae of 1947, their imaginary parts by Fermi's Golden Rule.

The details of the inductive construction sketched here are quite complicated; see [19]. For n large enough, it can be organized in the form of a "*renormalization group analysis*":

$$H^{(n+1)}(Z(z)) = \mathcal{R}\left(H^{(n)}(z)\right),$$

where $\mathcal{R}$ is a non-linear map (a "*renormalization map*") defined on a cylinder in a certain Banach space of (analytic families of) effective Hamiltonians on $\mathcal{F}^{(0)}$. Complex multiples of H_f turn out to be *fixed points* of $\mathcal{R}$. The map $\mathcal{R}$ has a stable manifold of co-dimension 2, an unstable manifold consisting of complex multiples of the identity and a one-dimensional manifold of fixed points, [19].

The strategy described here is implemented in [19]. It establishes a precise version of the conventional picture of an atom coupled to the quantized radiation field: It has a finitely degenerate groundstate. Excited states of the atom are *unstable* and decay by spontaneous emission of photons; given the results of [19,21], it is straightforward to estimate the *life times* of excited states, [22].

The picture, as of now partly conjectural, is that an arbitrary initial excited state of an atom coupled to the quantized electromagnetic field of finite total energy decays towards the *groundstate of the atom* accompanied by an *outgoing* flux of photons, through a cascade of intermediate, metastable states. As far as locating resonances is concerned, this picture is made rigorous in [19] (under the simplifying assumptions specified above). The remaining job is to translate the *time-independent results* of [19] into results describing the evolution of the system in *space and time*.

The problems and results described here are among the first hard results in an area of quantum physics of central importance for the *interpretation of the theory* and for understanding *classical regimes* in quantum theory: One studies a "small", localized system (e.g. an electron or an atom) coupled to an infinitely extended, *dispersive* quantum-mechanical medium (e.g. the radiation field); one chooses to only follow the time evolution of a finite number of "observables" of the small system. Because the small system is coupled to a medium of infinitely many degrees of freedom, and because this medium is dispersive, the time evolution of *some* of the observables of the *small system* can turn out to be *dissipative* and may, asymptotically, be well approximated by solutions of certain *classical* equations of motion. An example of such an *"open system"* for which the scenario just described can be made quite precise is a *laser* described within a mean field approximation, as discussed by K. Hepp and E. H. Lieb [23]. (For results valid within certain approximations, such as the van Hove limit, see also [24]). Rather concrete results for fairly general classes of systems have recently been established by Jaksic and Pillet [25].

Roughly thirteen years ago, Tom Spencer and I attempted to study such systems and considered the Heisenberg equations of motion within a Hartree-type approximation [26]. This led us to analyze certain non-linear wave- and Schrödinger equations. The properties of solutions of these equations illustrate the claim that, in classical Hamiltonian systems, or in unitary quantum-mechanical systems, *dissipation ("friction") accompanied by dispersion* is quite abundant. Precise results appeared in work of Sigal [27] and of Soffer and Weinstein [28].

General methods to study friction in classical and quantum mechanics remain to be developed; but some heuristic understanding of the mechanisms has been gained. See also [25].

3.4. Many-body theory: electron liquids, superconductivity, etc.
Among the most important goals of non-relativistic quantum theory is to understand properties of very large quantum systems, such as atomic gases [17, 5], liquids, and solids (e.g. insulators, normal metals, or superconductors). The results reviewed in Section 2 show that such systems exhibit a *fundamental gauge invariance* (under local rotations in spin space and local phase rotations). This gauge invariance turns out to have surprising consequences for a general classification of states of electron gases at positive density and low temperatures and in understanding a variety of general effects in condensed matter physics; see [29] for a systematic review. However, for the analysis of concrete systems, general principles are insufficient, and one must put to work tools of hard analysis. The front runner among such tools is (as in Section 3.3) the renormalization group method, in the form developed by Wilson [15]. A systematic renormalization group analysis of electron liquids at positive density was initiated by Feldman, Trubowitz and their coworkers [30], roughly ten years ago. Their work aims at mathematical rigor; it is still ongoing, at present. A description of their methods and results lies beyond the scope of this brief review. A summary of heuristic aspects of their work, with emphasis of connections to some models of relativistic Fermi fields in two space-time dimensions, may be found in [29].

Condensed matter physics and non-relativistic many-body theory are full of interesting and deep problems of analysis. As an example we mention the study of renormalization flows in general electron liquids which makes contact with very interesting problems in dynamical systems theory; (e.g. study of flows generated by vector fields on $\mathbb{R}^n$, $n \geq 3$, with "quadratic zeroes").

4. Quantum Hall effect, ballistic wires, and number theory

A feature that makes mathematical physics so fascinating is that it sometimes reveals surprising mathematical connections between seemingly unrelated phenomena. One example is the connection between certain quantum field theories of scalar fields and the physics of polymer chains, first found by Symanzik, which led to a variety of "non-interaction theorems" [31] that, intuitively, reflect the fact that two Brownian paths starting at distinct points of $\mathbb{E}^n$ miss each other with probability 1 if $n \geq 4$.

Another example is the connections between topological Chern-Simons theory in three dimensions, Kac-Moody (current) algebras, the theory of critical phenomena in two dimensions, the theory of knots and links in three-dimensional manifolds, the fractional quantum Hall effect, the theory of ballistic (quantum) wires and other one-dimensional quantum chains, the theory of integral odd lattices, etc.; see e.g. [32] and refs. given there (in particular [33, 29]). It would lead too far to describe these connections in detail. But it may be instructive to sketch some ideas related to the fractional quantum Hall effect and ballistic wires and see how they lead to problems concerning integral, odd lattices. [The precise physical situations we have in mind cannot be described here; but see [29] and references given there.]

We consider a gas of electrons confined to a two-dimensional layer approximately corresponding to an annulus Ω contained in a plane of physical space $\mathbb{E}^3$. A magnetic field $\vec{B}_0$ is turned on in a direction perpendicular to Ω. We choose coordinates x^1, x^2, x^3 in $\mathbb{E}^3$ such that Ω is contained in the plane $x^3 = 0$, and $\vec{B}_0 = (0, 0, B_0)$. Let $\vec{X}$ denote a vector field on $\mathbb{E}^3$. We decompose $\vec{X}\big|_\Omega$ into components parallel and normal to $T.\,\Omega$:

$$(4.1) \qquad \vec{X} = (\underline{X}, X), \quad \underline{X} := (X^1, X^2), \quad X := X^3.$$

Given $\underline{X}$, we define $\widetilde{\underline{X}} := (X^2, -X^1)$.

We propose to study the response of the electron gas in Ω to turning on a small, external electric field $\underline{E}$. It is assumed that the electrons cannot escape from Ω. Let $\underline{j}$ denote the electric current density in Ω, and let ρ denote the electric charge density. Experimentally, one finds that

$$(4.2) \qquad \underline{j} = \sigma_L \underline{E} + \sigma_H \widetilde{\underline{E}},$$

where σ_L is the longitudinal- and σ_H is the *Hall conductivity*. We consider an "*incompressible*" electron gas for which $\sigma_L = 0$. The Hall conductivity is then piecewise constant, and we assume that σ_H is equal to some positive constant on Ω and vanishes on $\{x^3 = 0\}\backslash\Omega$.

Since electric charge is conserved, and because electrons cannot escape from Ω, ρ and $\underline{j}$ satisfy the continuity equation

$$(4.3) \qquad \frac{\partial}{\partial t}\rho + \underline{\nabla} \cdot \underline{j} = 0,$$

where t denotes time.

Let $B_{\text{tot}} = B_0 + B$ denote the total magnetic field perpendicular to Ω; (B is a small perturbation of the constant field B_0). Then Faraday's induction law says that

$$(4.4) \qquad \frac{\partial}{\partial t}B_{\text{tot}} + \underline{\nabla} \cdot \widetilde{\underline{E}} = 0.$$

Combining eqs. (4.2), for $\sigma_L = 0$, (4.3) and (4.4), we find that

$$(4.5) \qquad \frac{\partial}{\partial t} B_{\text{tot}} = -\underline{\nabla} \cdot \underline{\tilde{E}} = -\sigma_H^{-1} \underline{\nabla} \cdot \underline{j} = \sigma_H^{-1} \frac{\partial}{\partial t} \rho.$$

Let ρ_0 denote the electric charge density (constant on Ω) of the system when $B = 0$, $\underline{E} = 0$, and let $j^0 := \rho - \rho_0$. Integrating (4.5) in time (with initial conditions $B = 0$, $j^0 = 0$), we find that

$$(4.6) \qquad j^0 = \sigma_H B.$$

Let $\Lambda := \mathbb{R} \times \Omega$ denote the *space-time* of the electron gas; $x^0 := t$. Let J denote the 2-form dual to the vector field $(j^0, \underline{j})$ on Λ, and let $F = \frac{1}{2} F_{\mu\nu} \, dx^\mu \wedge dx^\nu$, $\mu, \nu = 0, 1, 2$, denote the electromagnetic field tensor (2-form) on Λ, with

$$(F_{\mu\nu}) := \begin{pmatrix} 0 & E^1 & E^2 \\ -E^1 & 0 & -B \\ -E^2 & B & 0 \end{pmatrix}.$$

Then eqs. (4.2), for $\sigma_L = 0$, (4.3), (4.4) and (4.6) read

$$(4.7) \qquad J = \sigma_H F$$

$$(4.8) \qquad dJ = 0, \quad dF = 0.$$

Eqs. (4.8) are integrated by introducing 1-forms b and A (the electromagnetic vector potential) such that

$$(4.9) \qquad J = db, \quad F = dA.$$

Extending the definitions of b, A and σ_H from Λ to all of $\mathbb{R}^3$ by setting σ_H to zero, outside Λ, we can study eqs. (4.7) and (4.8) on $\mathbb{R}^3$. We recall that electrons cannot escape from Ω, which implies that $J\big|_{\mathbb{R}^3 \setminus \Lambda} = 0$ and that the component of $\underline{j}\big|_{\partial\Omega}$ normal to $\partial\Omega$ vanishes, (at all times). We now find that eqs. (4.7) and (4.8) are *incompatible*, unless J is replaced by $J_{\text{tot}} := J + J_\partial$, where J_∂ is a distributional 2-form (de Rham current) with support on $\partial\Lambda$ dual to a distributional vector field I supported on $\partial\Lambda$ and everywhere parallel to $\partial\Lambda$. The distribution J_∂ describes the famous *edge currents* observed in such two-dimensional electron gases. Eqs. (4.7) and (4.8) imply that

$$(4.10) \qquad dJ_\partial = d\sigma_H \wedge F.$$

If E_θ denotes the component of $\underline{E}\big|_{\partial\Omega}$ parallel to $\partial\Omega$, we find that (if I_θ is constant in θ)

$$(4.11) \qquad \frac{\partial}{\partial t} I^0 = \sigma_H E_\theta$$

on $\partial\Lambda$. We now return to studying eqs. (4.7) and (4.9) valid in the interior of Λ. They yield the equations

$$(4.12) \qquad db = \sigma_H \, dA.$$

These equations are the Euler-Lagrange variational equations for the action functional

$$(4.13) \qquad S_\Lambda(b; A) = \frac{1}{2} \sigma_H^{-1} \int_\Lambda b \wedge db + \int_\Lambda db \wedge A$$

under variations of the *dynamical* variable b. The functional S_Λ is the *abelian topological Chern-Simons action.* It does not depend on the edge current J_∂. In order to describe the dynamics of edge currents, a boundary term must be added to S_Λ.

The current densities J and J_∂ must be interpreted as expectation values of quantum-mechanical, operator-valued distributions, $\mathcal{J}$ and $\mathcal{J}_\partial$, in suitable states of the system. The quantum theory of $\mathcal{J}$ is quantized Chern-Simons theory: Interpreting b quantum-mechanically, and setting $\mathcal{J} = db$, the action functional in a *path integral formulation* of the quantum theory of $\mathcal{J}$ is given by (4.13). This quantum theory is known to be "topological" [33]; (the Hamiltonian vanishes). But there is a problem with the action functional (4.13): It violates *gauge invariance*,

$$(4.14) \qquad\qquad A \mapsto A + d\alpha, \quad b \mapsto b + d\beta,$$

where α and β are functions on Λ not necessarily vanishing at the boundary $\partial\Lambda$. Under the transformations (4.14), the action functional $S_\Lambda(b; A)$ changes by a *boundary term*, (an integral of a 2-form over $\partial\Lambda$). This lack of gauge invariance is not surprising, because, in (4.13), we have omitted a boundary term describing the dynamics of the edge current, $\mathcal{J}_\partial$, coupled to the external electromagnetic vector potential A. The dynamics of *chiral* edge currents violates electromagnetic gauge invariance (invariance under $A \mapsto A + d\alpha$), as well! This violation of gauge invariance, the "two-dimensional, chiral anomaly" [34], precisely cancels the one exhibited by the Chern-Simons theory of the bulk current $\mathcal{J}$. Thus, the dynamics of the *entire* system of bulk currents ($\mathcal{J}$) and chiral edge currents ($\mathcal{J}_\partial$) does *not* violate electromagnetic gauge invariance, (as required by fundamental principles of quantum theory).

Because the quantum theory of the bulk currents is topological, the *dynamical* degrees of freedom of a two-dimensional (2D), incompressible ($\sigma_L = 0$) electron gas in a transverse magnetic field are described by the edge current, $\mathcal{J}_\partial = \mathcal{J}_{\partial_+} + \mathcal{J}_{\partial_-}$, where $\mathcal{J}_{\partial_+}$ and $\mathcal{J}_{\partial_-}$ are the edge currents localized at the two components, C_+ and C_-, of the boundary of the annulus Ω to which the 2D electron gas is confined. The edge currents $\mathcal{J}_{\partial_+}$ and $\mathcal{J}_{\partial_-}$ have *opposite* chiralities. The operator-valued distribution $\mathcal{J}_{\partial_\pm}$ is the generator of a chiral U(1) Kac-Moody algebra contained in a chiral algebra $\mathcal{G}_\pm \simeq \mathcal{G}$. Besides the Kac-Moody algebra, this algebra contains a Virasoro algebra reflecting the "conformal invariance" of the quantum theory of edge currents.

Experimentally, the Hall conductivity σ_H is measured as follows: The degrees of freedom localized at the two components, C_+ and C_-, of the boundary of the annulus Ω to which the electron gas is confined are coupled to two electron reservoirs with slightly different chemical potentials, μ_+ and μ_-; (one may imagine that C_+ is connected to one lead and C_- to the other lead of a battery). The potential drop, V, between electrons near C_+ and electrons near C_- is then given by

$$e V = \mu_+ - \mu_-.$$

One measures the *total edge current*,

$$(4.15) \qquad\qquad I = \langle \mathcal{J}_{\partial_+} \rangle_{\mu_+} + \langle \mathcal{J}_{\partial_-} \rangle_{\mu_-},$$

i.e., the sum of the edge currents carried by states $\langle\,(\,\cdot\,)\,\rangle_{\mu_\pm}$ of the degrees of freedom localized near $C_\pm$; (in (4.15), $\mathcal{J}_{\partial_\pm}$ is identified with the θ-component of the edge current). The experimental value of the Hall conductivity σ_H is determined by measuring V and I and setting

(4.16) $\sigma_H = I/V.$

One important task for theorists is to predict the possible values of σ_H, assuming that $\sigma_L = 0$. It is in studying this problem that one is led to consider integral odd lattices.

Let $N_+ = N_- = N$ denote the central charge of the Virasoro algebras contained in the chiral algebras $\mathcal{G}_+ \simeq \mathcal{G}_- \simeq \mathcal{G}$. In the "minimal models" of 2D incompressible electron gases, N turns out to be a positive integer. It is worthwhile to remark that, in principle, N can be measured in *heat conduction experiments*. This integer is important in the analysis of the representation theory of the chiral algebra $\mathcal{G}$: In the "minimal model" [37], the unitary highest-weight representations of $\mathcal{G}$ can be labeled by elements, q, of an N-dimensional, real vector space, $\mathbb{R}^N$, which is equipped with a quadratic form $\langle \cdot, \cdot \rangle$. To every representation $q \in \mathbb{R}^N$ of $\mathcal{G}$ one can assign its *quantum statistics* (Bose-, Fermi-, or fractional statistics), which is given by $\langle q, q \rangle \bmod 2$: if $\langle q, q \rangle \equiv 0 \pmod 2$ then q has Bose statistics, $\langle q, q \rangle \equiv 1 \pmod 2$ corresponds to Fermi statistics, and $\langle q, q \rangle \neq 0, 1 \pmod 2$ means that q has fractional statistics. Furthermore, to every representation q of $\mathcal{G}$ one can assign its *electric charge* $Q(q)$ (in units where the elementary electric charge $e = 1$), where Q is some linear functional on $\mathbb{R}^N$. Among the representations of $\mathcal{G}$ that can be realized, experimentally, in a two-dimensional electron gas, there are all those representations $q \in \mathbb{R}^N$ corresponding to an integer electric charge,

(4.17) $Q(q) \in \mathbb{Z}.$

The quantum theory of electron gases imposes the constraint that if q is a physical (i.e., experimentally realizable) representation of $\mathcal{G}$ of integer electric charge then

(4.18) $Q(q) \equiv \langle q, q \rangle \bmod 2.$

Moreover, if q is a physical representation of $\mathcal{G}$ then so is $-q$, and if q_1 and q_2 are physical representations of $\mathcal{G}$ then so is $q_1 + q_2$.

It follows that the physical representations of $\mathcal{G}$ of integer electric charge can be identified with the sites of an N-*dimensional, integral, odd lattice* Γ, and the functional Q can be identified with a "visible" lattice point in the *dual lattice* Γ^*; see [35].

The pair (Γ, Q) of an N-dimensional, integral odd lattice and a point $Q \in \Gamma^*$ depends on the specific two-dimensional (2D), incompressible ($\sigma_L = 0$) electron gas that one considers. The task of associating a specific pair (Γ, Q) with a given 2D incompressible electron gas is a very hard problem in the analysis of non-relativistic quantum theory. It is clear that *not* every pair (Γ, Q) corresponds to a physically realizable 2D incompressible electron gas! [The special case, where Γ is the N-dimensional, Euclidian, simple lattice $\mathbb{Z}^N$, and $Q = (1, \ldots, 1) \in (\mathbb{Z}^N)^* \simeq \mathbb{Z}^N$, describes a system of *non-interacting, spin-polarized* electrons with N filled Landau levels. This special case is the only one that is completely understood, mathematically. It describes the *integer* quantum Hall effect.] There are fairly precise mathematical results, derived from phenomenologically justified assumptions concerning properties of 2D incompressible electron gases, which describe the class of all those pairs (Γ, Q) that might correspond to an experimentally realizable, 2D incompressible electron gas; see [36, 29]. These results are based on restricting the possible values of certain numerical invariants of (Γ, Q) to physically relevant ranges.

For the theoretical physicist, one of the amusing parts of the analysis is to show that, under the experimental conditions described above and starting from eq. (4.16), one is led to the formula

$$(4.19) \qquad\qquad \sigma_H = \frac{e^2}{h} \langle Q, Q \rangle,$$

where e is the elementary electric charge and h is Planck's constant; see [29], and [37] for a new derivation. Because Γ is an integral lattice and $Q \in \Gamma^*$, it follows that $\langle Q, Q \rangle$ is a *rational* number. The term "fractional quantum Hall effect" refers to the property of σ_H to be a *rational multiple* of a fundamental constant of nature, $\frac{e^2}{h}$, provided $\sigma_L = 0$ (a property first discovered experimentally; see [38]).

Among physically important *invariants* of (Γ, Q), there are the dimension and the discriminant, $|\Gamma^*/\Gamma|$, of Γ, the Kneser shape of Γ, the genus of Γ, the "Hall fraction" $\langle Q, Q \rangle$, as well as a number of invariants that appear to have escaped the attention of mathematicians. One of the interesting mathematical problems to be dealt with is to enumerate or classify all those pairs (Γ, Q) with the property that the values of certain invariants, such as those mentioned above, when evaluated for (Γ, Q) belong to a physically relevant range. For every pair (Γ, Q) in this family, one computes $\sigma_H = \frac{e^2}{h} \langle Q, Q \rangle$. The agreement between theoretical predictions and experimental observations turns out to be very satisfactory. For concrete results see [29, 35, 36].

Although all details have to be omitted here, I hope that the brief outline of the theory of the fractional quantum Hall effect given above illustrates the following point: It can happen that one starts with a dirty-looking problem in non-relativistic, quantum-mechanical many-body theory and, after a sometimes rather long reasoning process, one ends up with some beautifully clear problems in pure mathematics, e.g. in number theory. It is such features of mathematical physics that make it such an exciting area to work in.

To conclude, I should mention that the theory of conductance quantization for electron transport through very pure ("ballistic") wires [39] is very similar to the one outlined above for 2D incompressible electron gases; see [40]. The conductance is given by the right-hand side $\left(\frac{e^2}{h}\langle Q, Q \rangle\right)$ of eq. (4.17). But it appears that, in the theory of ballistic wires, only the pairs

$$(4.20) \qquad\qquad \Gamma = \mathbb{Z}^N, \quad Q = (1, \ldots, 1), \quad N \text{ even},$$

are physically realizable. The evenness of N is a direct experimental signature of *electron spin*.

The physics of ballistic wires is interesting as an example of *transport in thermal equilibrium* and of the role of *conformal invariance* in the quantum theory of one-dimensional systems.

Readers interested in further aspects of the problems described in this section are referred to [41, 29, 32, 35, 36].

5. Quantum theory of space-time and non-commutative geometry

In Section 2, we considered the non-relativistic quantum theory of a point particle of spin $\frac{1}{2}$ moving in a physical space described by a Riemannian spin^c manifold (M, g). We noted that if the manifold M is *not* a spin manifold but admits a spin^c structure then the quantum theory of a spin $\frac{1}{2}$-particle moving on M can be formulated consistently only if the particle is *electrically charged*, so that parallel transport on the spinor bundle S can be defined in terms of a spin^c connection ∇^S (see eq. (2.4)) that depends on a "virtual U(1)-connection" $\frac{1}{2}A$. The physical interpretation of A is that it is the *electromagnetic vector potential* describing an external magnetic field; and the point particle we are talking about is an electron.

In Section 3.1 (eqs. (3.12)–(3.18)) and in Section 3.3 (eqs. (3.24), (3.25)), we noted that the electromagnetic field must be *quantized* if we want to reach agreement between theory and experiment. The vector potential A must be interpreted as an operator-valued distribution with a non-trivial dynamics of its own!

Reasonable people would agree that if *some* of the components of a spin^c connection ∇^S, the ones depending on A, must be viewed as *quantum-mechanical*, operator-valued distributions then it is likely that *all* the components of ∇^S are *dynamical, quantum-mechanical*, operator-valued distributions. If ∇^S is compatible with the Levi-Civita connection, ∇, on $\Omega^{\cdot}(M)$ then it is determined by the metric, g, on M and the electromagnetic vector potential A. We are thus tempted to conclude that, in physics, the metric g, too, must be treated as a *dynamical, quantum-mechanical* object. This is the fundamental problem of "quantum gravity".

Classical, relativistic physics at large distance- and low-energy scales instructs us to unify space and time to a Lorentzian space-time (N, η), where N is an $(n+1)$-dimensional manifold, and η is a Lorentzian metric on TN describing the gravitational field. [Space-time is assumed to have certain good properties: N should admit a global causal orientation; Gödel universes should be excluded, and the "cosmic censorship hypothesis" should be valid.] In classical, relativistic physics at large distance- and low-energy scales, the dynamics of η is determined by solving *Einstein's field equations*

$$(5.1) \qquad\qquad G(\eta) = l_P^{n-2} T,$$

where $G(\eta) := \mathrm{Ricci}\,(\eta) - \frac{1}{2}\,g\,r(\eta)$ is the Einstein tensor, and T is the energy momentum tensor of matter; furthermore l_P, the *Planck length*, is a constant of nature with the dimension of length. In our world, $n = 4$ and $l_P \approx 10^{-33}$cm (corresponding to an energy of $\approx 10^{19}$ GeV; l_P can be calculated from Newton's gravitational constant, Planck's constant and the velocity of light).

Matter is quantum-mechanical, and T is therefore an operator-valued distribution on a Hilbert space (albeit seemingly an ill-defined one). Hence Einstein's equations appear to tell us that $G(\eta)$, too, must be an operator-valued distribution. This, unfortunately, causes problems with the usual interpretation of space-time as a *classical* Lorentzian manifold, (N, η). One should ask, therefore, what kind of mathematical structure describes space-time, at a fundamental level? It is likely that a better question to ask is: What kind of mathematical structure describes *space-time-matter*? Finding the answer to this question will plausibly remain one of the most fundamental problems of physics, for many years to come; and it will trigger plenty of mathematical activity.

Rather than immediately trying to find the answer, one may try to guess some general features of the answer. We may, for example, engage in a bold physical speculation, to the effect that quantum theory and the general theory of relativity remain *qualitatively* correct up to energy scales of the order of 10^{19} GeV. Of course,

this speculation may turn out to be wrong. But assuming that it is correct we arrive
at some interesting general conclusions: A localized, observable event in space-time
is always the radiative decay of a localized, unstable state of a quantum-mechanical
(sub-)system. Let us assume that the event has been observed during a time interval
of length Δt, (Δt is the life time of the unstable state). Imagine, furthermore, that,
by observing the decay products (radiation), we are able to localize the event within
a spatial region of maximal extension d' and minimal extension d''. By combining
Heisenberg's uncertainty relations, Hawking's laws of black hole evaporation and
Einstein's equations (5.1), one can argue that Δt, d' and d'' obey the following
uncertainty relations

$$(5.2) \qquad \Delta t \cdot d' \geq l_P^2, \quad d' \cdot d'' \geq l_P^2$$

formulated in [42]; (see also [3] for further discussion). One can argue, furthermore,
that the number of observable events in a *compact* region of space-time is *finite*.

The most plausible conclusion from this discussion is to say that with *real* mi-
croscopes one *cannot* resolve space-time regions of diameter $\ll l_P$. This suggests
that the notion of a classical space-time continuum is *not* strictly meaningful in
physics, and that, ultimately, space-time has quantum-mechanical features, i.e., is
described by a non-commutative operator algebra. [This idea is somewhat analogous to the one that describes the passage from classical, Hamiltonian mechanics to
quantum mechanics as a deformation quantization of classical phase space, which
is the subject of geometric quantization.] It is, a priori, very unclear how one
should proceed to quantize space-time (-matter), and several approaches have been
suggested. One important result of [42] is to show that the relations (5.2) are
compatible with relativistic (Poincaré) invariance of space-time physics. This suggests that a quantum theory of space-time-matter may incorporate some features
of relativistic physics.

At present, the most likely candidates for a quantum theory of space-time-
matter are "theories" of *extended objects*, such as superstring- [43] and M-theory.
Such "theories" automatically incorporate uncertainty relations of the form (5.2). If
we are lucky we shall learn how to formulate these theories *without* reference to any
specific model of space-time. They will then, a priori, not talk about space-time, and
space-time should emerge as a *derived* structure. [An analogy may be the emergence
of classical dynamics in certain regimes of open, quantum-mechanical systems, as
mentioned in Section 3.3.] One idea of how to rediscover (quantum) space-time in
superstring theory is to study the geometry of *superstring vacua*, viewed as certain
superconformal field theories. It turns out that some superconformal field theories
describing aspects of superstring vacua provide one with *supersymmetric spectral
data*, $(\mathcal{A}, D, \mathcal{H})$ or $(\mathcal{A}, \mathcal{D}, \overline{\mathcal{D}}, *, \mathcal{H})$, etc., of the type discussed in Section 2 (in the
context of the non-relativistic quantum theory of an electron or of positronium),
but with the difference that $\mathcal{A}$ is usually a *non-commutative* *algebra. One is led to
interpret $\mathcal{A}$ as an "algebra of functions" on a quantized phase space associated to the
loop space over some compact, generally non-commutative geometrical space M.
[The space M corresponds to an internal space for a superstring vacuum describing a
quantum space-time of the form $N \simeq \mathbb{M}_q^d \times M$, where $\mathbb{M}_q^d$ is a quantum deformation
of, for example, d-dimensional Minkowski space.] The problem arises of how to
reconstruct the space M and its non-commutative differential geometry; see [44, 3].
This is a difficult problem in conformal field theory and non-commutative geometry.
It has been solved for some toy examples, [44, 3]. As a special case, we mention

the examples where M is a quantum deformation of a *group manifold* (associated with a compact, classical Lie group). Generally speaking, the task is to extract from $(\mathcal{A}, D, \mathcal{H})$ (or $(\mathcal{A}, \mathcal{D}, \overline{\mathcal{D}}, *, \mathcal{H}), \dots$) supersymmetric spectral data $(\mathcal{A}_0, D_0, \mathcal{H}_0)$ (or $(\mathcal{A}_0, \mathcal{D}_0, \overline{\mathcal{D}}_0, *, \mathcal{H}_0), \dots$) describing the differential geometry of the space M. In particular, the algebra $\mathcal{A}_0$ is supposed to be a (generally non-commutative) "algebra of functions" on M. The construction of $(\mathcal{A}_0, D_0, \mathcal{H}), \dots$, from the spectral data $(\mathcal{A}, D, \mathcal{H}), \dots$, provided by a superconformal field theory is the difficult step. In particular, the reconstruction of the correct algebra $\mathcal{A}_0$ from a superconformal field theory remains somewhat mysterious, in general.

But once supersymmetric spectral data $(\mathcal{A}_0, D_0, \mathcal{H}_0)$ (or $(\mathcal{A}_0, \mathcal{D}_0, \overline{\mathcal{D}}_0 d, *, \mathcal{H}_0)$, etc.) are given, it is quite well understood how to reconstruct a non-commutative geometrical space, M, from such data and how to explore its differential geometry. The relevant mathematical techniques have been pioneered by Connes; see [1]. Inspired by the non-relativistic quantum theory of the electron and of positronium and motivated by the results of [4], we have cast them in a form particularly useful for applications to quantum theory; see [2, 3]. It is not possible to enter into a description of precise mathematical results, because this would take some space. I therefore refer the reader to the literature; in particular to [1, 2, 3, 4].

The lesson is that an approach to differential geometry based on the supersymmetric, non-relativistic quantum theory of an electron and of positronium, combined with operator algebra theory, guides a way to far-reaching generalizations of differential geometry, in particular to *non-commutative* differential geometry. An exploration of the geometrical foundations of the quantum theory of electrons interacting with the quantized electromagnetic field suggests that, ultimately, space-time, too, is quantum-mechanical. This is the topic of quantum gravity. Quantum gravity and non-commutative geometry will most likely be fellow-travellers.

It seems clear that the electron will remain inexhaustible for many years to come!

In conclusion, I wish to thank my collaborators and Ph. D. students for the joy I found in our joint efforts.

References

[1] A. Connes, "Noncommutative Geometry", Academic Press, New York 1994.

[2] J. Fröhlich, O. Grandjean and A. Recknagel, "Supersymmetric Quantum Theory and Differential Geometry", "Supersymmetric Quantum Theory and Non-Commutative Geometry", to appear in Commun. Math. Phys.

[3] J. Fröhlich, O. Grandjean and A. Recknagel, "Supersymmetric Quantum Theory, Non-Commutative Geometry, and Gravitation", to appear in the proceedings of the 1995 Les Houches summer school on "Quantum Symmetries", A. Connes, K. Gawędzki and J. Zinn-Justin (eds.).

[4] E. Witten, Nucl. Phys. B **202**, 253–316 (1982); J. Diff. Geom. **17**, 661–692 (1982). [See also L. Alvarez-Gaumé, Commun. Math. Phys. **90**, 161–173 (1983).]

[5] E. H. Lieb, "The Stability of Matter: From Atoms to Stars", Selecta, 2nd edition, Springer-Verlag, Berlin, Heidelberg, New York 1997.

[6] W. Hunziker and I. M. Sigal, "The General Theory of n-Body Quantum Systems", in: "Mathematical Quantum Theory II, Schrödinger Operators", J. Feldman et al. (eds.), AMS Publ., Montreal 1994.

[7] J. Fröhlich, E. H. Lieb and M. Loss, Commun. Math. Phys. **104**, 251–270 (1986).

[8] E. H. Lieb and M. Loss, Commun. Math. Phys. **104**, 271–282 (1986).

[9] M. Loss and H.-T. Yau, Commun. Math. Phys. **104**, 283–290 (1986).

[10] N. Seiberg and E. Witten, Nucl. Phys. B **426**, 19–52 (1994).
      E. Witten, "Monopoles and 4-Manifolds".
[11] C. Fefferman, Proc. Natl. Acad. Sci. **92**, 5006–5007 (1995).
[12] E. H. Lieb, M. Loss and J. P. Solovej, Phys. Rev. Lett. **75**, 985–989 (1995).
[13] C. Fefferman, J. Fröhlich and G. M. Graf, "Stability of Ultraviolet-Cutoff Quantum Electro-
      dynamics with Non-Relativistic Matter", to appear in Commun. Math. Phys.;
      C. Fefferman, "On Electrons and Nuclei in a Magnetic Field", to appear in Adv. Math. .
[14] L. Bugliaro-Goggia, J. Fröhlich and G. M. Graf, Phys. Rev. Lett. **77**, 3494–3497 (1996).
[15] K. G. Wilson and J. Kogut, Phys. Reports **12**, 75–200 (1974).
[16] T. Chen and J. Fröhlich, unpublished notes (1997).
[17] C. Fefferman, Rev. Math. Iberoamericana **1**, 1–44 (1985).
      J. Conlon, E. H. Lieb and H.-T. Yau, Commun. Math. Phys. **125**, 153–180 (1989).
[18] J. Fröhlich, G. M. Graf, J.-M. Richard and M. Seifert, Phys. Rev. Lett. **71**, 1332– (1993).
      M. Seifert, Ph. D. thesis, ETH-Zürich, 1997.
[19] V. Bach, J. Fröhlich and I. M. Sigal, Lett. Math. Phys. **34**, 183–201 (1995);
      "Quantum Electrodynamics of Confined Non-Relativistic Particles", to appear in Adv. Math.;
      "Renormalization Group Analysis of Spectral Problems in Quantum Field Theory", to appear
      in Adv. Math.
[20] T. Kato, "Perturbation Theory for Linear Operators", Springer-Verlag, Berlin, Heidelberg,
      New York, 1980.
[21] W. Hunziker, Commun. Math. Phys. **132**, 177–188 (1990).
[22] T. Chen, diploma thesis, ETH-Zürich 1996; (see also ref. 16).
[23] K. Hepp and E. H. Lieb, Ann. Phys. (NY) **76**, 360–404, (1973); Helv. Phys. Acta **76**, 573–602
      (1973); "Constructive Macroscopic Quantum Electrodynamics", in "Constructive Quantum
      Field Theory", Proceedings of the 1973 Erice Summer School, G. Velo and A. S. Wightman
      (eds.), Springer Lecture Notes in Physics, vol. 25, 298–316 (1973).
[24] E. B. Davies, "Quantum Theory of Open Systems", Academic Press, London and New York,
      1976.
[25] V. Jaksic and C. A. Pillet, Ann. Inst. H. Poincaré **62**, 47– (1995); Commun. Math. Phys.
[26] J. Fröhlich, T. Spencer and C. E. Wayne, J. Stat. Phys. **42**, 247–274 (1986).
      C. A. Pillet, "Mécanique Quantique dans un Potentiel Aléatoire Markovien", Ph. D. Thesis
      ETH 1986.
      C. Albanese and J. Fröhlich, Commun. Math. Phys. **116**, 475–502 (1988).
      C. Albanese, J. Fröhlich and T. Spencer, Commun. Math. Phys. **119**, 677–699 (1988).
      G. C. Benettin, J. Fröhlich and A. Giorgilli, Commun. Math. Phys. **119**, 95–108 (1988).
[27] I. M. Sigal, Commun. Math. Phys. **153**, 297–320 (1993).
[28] A. Soffer and M. Weinstein, Commun. Math. Phys. **133**, 119–146 (1990); J. Diff. Eqs. **98**,
      376–390 (1992); preprints to appear.
[29] J. Fröhlich, U. M. Studer and E. Thiran, "Quantum Theory of Large Systems of Non-
      Relativistic Matter", in: "Fluctuating Geometries in Statistical Mechanics and Field Theory",
      F. David, P. Ginsparg and J. Zinn-Justin (eds.), Elsevier, Amsterdam, 1995.
[30] J. Feldman and E. Trubowitz, Helv. Phys. Acta **63** (1990) 156;
      G. Benfatto and G. Gallavotti, J. Stat. Phys. **59** (1991) 541.
      J. Feldman and E. Trubowitz, Helv. Phys. Acta **64** (1991) 214.
      J. Feldman, J. Magnen, V. Rivasseau and E. Trubowitz, Helv. Phys. Acta **65** (1992) 679.
      J. Feldman, J. Magnen, V. Rivasseau and E. Trubowitz, Europhys. Lett. **24** (1993) 437.
      J. Feldman, J. Magnen, V. Rivasseau and E. Trubowitz, Europhys. Lett. **24** (1993) 521.
      J. Feldman, J. Magnen, V. Rivasseau and E. Trubowitz, Fermionic Many-Body Models, in:
      "Mathematical Quantum Theory I: Field Theory and Many-Body Theory", J. Feldman, R.
      Froese and L. Rosen (eds.), CRM Proceedings and Lecture Notes, vol. 7, AMS Publ., 1994.
      J. Feldman, D. Lehmann, H. Knörrer and E. Trubowitz, Fermi Liquids in Two Space Di-
      mensions, in: "Constructive Physics", V. Rivasseau, ed., Lecture Notes in Physics, vol. 446
      (Springer-Verlag, Berlin, Heidelberg, New York, 1995).
[31] M. Aizenman, Phys. Rev. Lett. **47**, 1–4 (1981); Commun. Math. Phys. **86**, 1–48 (1982).
      J. Fröhlich, Nuclear Physics **B 200** [FS 4], 281–296 (1982).
      R. Fernandez, J. Fröhlich and A. Sokal, "Random Walks, Critical Phenomena and Triviality
      in Quantum Field Theory", Springer-Verlag, Berlin, Heidelberg, New York 1992.

[32] Research Group in Mathematical Physics, "The Fractional Quantum Hall Effect, Chern-Simons Theory, and Integral Lattices", in: Proc. of ICM'94, S. D. Chatterji (ed.), Basel, Boston, Berlin: Birkhäuser Verlag 1995.

[33] E. Witten, Commun. Math. Phys. **121**, 351 (1989).

[34] R. Jackiw and R. Rajaraman, Phys. Rev. Lett. **54**, 1219 (1985).
R. Jackiw, in "Current Algebra and Anomalies", S. B. Treiman, R. Jackiw, B. Zumino and E. Witten (eds.), World Scientific Publ., Singapore 1985.
H. Leutwyler, Helv. Phys. Acta **59**, 201 (1986).

[35] J. Fröhlich and E. Thiran, J. Stat. Phys. **76**, 209–283 (1994).

[36] J. Fröhlich, U. Studer and E. Thiran, J. Stat. Phys. **86**, 821–897 (1997).

[37] J. Fröhlich, Lectures at ETH, 1996/97 (unpublished).

[38] K. von Klitzing, G. Dorda and M. Pepper, Phys. Rev. Lett. **45**, 494 (1980).
D. C. Tsui, H. L. Störmer and A. C. Gossard, Phys. Rev. Lett. **48**, 1559 (1982).

[39] B. J. van Wees et al., Phys. Rev. Lett. **60**, 848 (1988).
S. Tarucha, T. Honda and T. Saku, Solid State Commun. **94**, 413 (1995).
A. Yacoby et al., Phys. Rev. Lett. **77**, 4612 (1996).

[40] A. Yu. Alekseev, V. V. Cheianov and J. Fröhlich, Phys. Rev. **B 54**, R 17 320 (1996); "Universality of Equilibrium One-Dimensional Transport from Gauge Invariance", submitted to Phys. Lett.

[41] R. B. Laughlin, Phys. Rev. **B 23**, 5632 (1981); Phys. Rev. Lett. **50**, 1395 (1983); Phys. Rev. **B 27**, 3383 (1983).
D. J. Thouless, M. Kohmoto, M. P. Nightingale and M. den Nijs, Phys. Rev. Lett. **49**, 405 (1982).
J. E. Avron, R. Seiler and B. Simon, Phys. Rev. Lett. **51**, 51 (1983); Phys. Rev. Lett. **65**, 2185 (1990).
J. E. Avron and R. Seiler, Phys. Rev. Lett. **54**, 259 (1985).
J. Bellissard, in "Localization in Disordered Systems" (Bad Schandau), W. Weller and P. Ziesche (eds.), Teubner, Leipzig, 1988.

[42] S. Doplicher, K. Fredenhagen and J. E. Roberts, Commun. Math. Phys. **172**, 187 (1995).

[43] M. B. Green, J. H. Schwarz and E. Witten, "Superstring Theory", volumes I and II, Cambridge University Press, Cambridge 1987.

[44] J. Fröhlich and K. Gawędzki, "Conformal Field Theory and the Geometry of Strings", in "Mathematical Quantum Theory", J. Feldman, R. Froese and L. Rosen (eds.), CRM Proceedings and Lecture Notes, vol. 7, 57–97, AMS Publ., 1994.

THEORETICAL PHYSICS, HPZ, ETH–HÖNGGERBERG, CH–8093, ZÜRICH, SWITZERLAND
E-mail address: `froehlich@itp.phys.ethz.ch`

Quantitative Homotopy Theory

Mikhael Gromov

I once attended lectures about cosmology by two topologists, great topologists, concerning the possible shape of the Universe (it was about 25 years ago), and I asked the first one whether the Universe was simply connected or not. When asked this question he said "It is clear that the Universe cannot be but simply connected, for non-simple connectedness would imply some high-scale periodicity, which is ridiculous." The other's talk was entitled, "Is the Universe simply connected?" When I told him what the first said, he responded, "Who cares, it's still a meaningful question, like it or not." What I have to say is not exactly related to this, but is motivated by the naive question of whether or not it makes sense to ask of something that it be simply connected. By sense, I mean physical, in the spirit of Aristotle—everyday physics. When we ask this question we want a "physically" meaningful answer. So, we consider the question this way: take a loop in the Universe, a reasonably short loop compared to the size of the Universe, say of no more than 10^{10} to 10^{12} light years long and ask if it is contractible. And, to be realistic, we pick a certain time, for example 10^{30} years, and ask if it is contractible within this time. So you are allowed to move the loop around, say at the speed of light, and try to determine whether or not it can be contracted within this time. The point is, even imagining our space to be some topological 3-sphere S^3, we can organize an innocuous enough metric on S^3 so that it takes more than 10^{30} years to contract certain loops in this sphere and in the course of contraction we need to stretch the loop to something like 10^{30} light years in size. So, if 10^{30} years is all the time you have, you conclude that the loop is not contractible and whether or not $\pi(S^3) = 0$ becomes a matter of opinion.

My point is that when you have a space of maps, like the space of maps of a circle into a compact three manifold, there is, in the homotopy theory, some extra structure coming from geometry as the one just illustrated. Namely, when we speak of homotopy, we try to keep track of the sizes of maps and of homotopies. Here we encounter new questions, some of which I am going to discuss.

Consider compact finite dimensional spaces X and Y, say finite polyhedra. Now, a finite polyhedron has essentially a unique piece-wise Euclidean metric, unique meaning that if you have two metrics on X, then (perhaps after a little wiggling) we can construct a bi-Lipschitz homeomorphism between the two. In particular, for a compact manifold, two metrics differ by a multiplicative constant, and from our point of view, which will be concerned with orders of magnitude, these are essentially the same.

1991 *Mathematics Subject Classification.* 57N65.

So we are given metric spaces X and Y and we study the space of continuous maps $\operatorname{Map}(X \to Y)$. A basic characteristic of a map $f : X \to Y$ is its dilation or *Lipschitz constant* which measures by how much f stretches (the curves in) X. Namely,

$$\operatorname{Lip}(f) = \sup_{x_1 \neq x_2} \frac{\operatorname{dist}\big(f(x_1), f(x_2)\big)}{\operatorname{dist}(x_1, x_2)}.$$

Now we want to understand the structure of this function $f \mapsto \operatorname{Lip}(f)$ for $f \in \operatorname{Map}(X \to Y)$ pretending we perfectly understand the topology of the space of maps. For example, let X be the sphere S^n so that the connected components of $\operatorname{Map}(X \to Y)$ are represented by the homotopy group $\pi_n(Y)$ and suppose this group is isomorphic to $\mathbb{Z}$. We represent each $h \in \pi_n(Y) = \mathbb{Z}$ by a map f of minimal dilation and think of this dilation $h \mapsto \inf_{[f]=h} \operatorname{Lip}(f)$ as a kind of norm $\|h\|$ on $\pi_n(Y)$.

Notice that the individual values $\|h\|$, $h \in \mathbb{Z}$, of this "norm" depend on our metrics on S^n and Y but the asymptotics for $h \to \infty$ are essentially the same for all commonly used metrics.

Example. Consider maps $f : S^2 \to S^2$. Every such f is characterized, up to homotopy, by the degree $d \in \mathbb{Z} = \pi_2(S^2)$. One easily constructs maps f of degree d and $\operatorname{Lip}(f) \leq 100\sqrt{d}$ for the standard metric on S^2 and all $d \in \mathbb{Z}$, thus showing $\|d\| = O(\sqrt{d})$ for $d \to \infty$ and for *all* Riemannian metrics on S^2. On the other hand, $\|d\|$ can not be much smaller than $\sqrt{d}$ since

$$d = \deg(f) = \int_{S^2} \operatorname{Jac}(f)\, ds \,,$$

where the Jacobian of f is quadratic in the partial derivatives of f and thus is bounded by $\big(\operatorname{Lip}(f)\big)^2$. Therefore,

$$\|d\| \sim \sqrt{d}\,.$$

Generalization. Take an $h \in \pi_n(Y)$ such that the *Hurewicz homomorphism* does not vanish on h, not even after tensoring with $\mathbb{R}$. Then, our "Lipschitz norm" of the powers $h^d \in \pi_n(X)$ grows as $\operatorname{const} d^{\frac{1}{n}}$, where the upper bound on $\|h^d\|$ is obtained with easy maps $S^n \to S^n$ of degree d and dilation $\leq C_n d^n$, while the lower bound appeals to the volume growth of maps f_d representing h^d, or, equivalently to the growth of the integrals $\int_{S^n} f^*(\omega)$ for some closed n-form ω on Y for which $\int_{S^n} f_d^*(\omega) \neq 0$.

Hopf maps. This map, $f : S^3 \to S^2$ is homologous to zero (i.e. Hurewicz $[f] = 0$) as $H_3(S^2) = 0$, yet it represents a *nontrivial* class $h \in \pi_2(S^3) = \mathbb{Z}$. The self-mappings $\varphi : S^2 \to S^2$ of degree k transform this h to h^{k^2} which easily shows that the norm $\|h^d\|$ grows no faster than $\operatorname{const} d^{\frac{1}{4}}$ (for all d, not only for those of the form k^2). And the lower bound follows from the following definition of the Hopf invariant $h(f) \in \mathbb{Z} = \pi_3(S^2)$ due to Whitehead. Take the area form ω on S^2 and let ω'_f be a primitive (1-form) of the pull-back $\omega_f^* = f^*(\omega)$ on S^3, i.e. $d\omega'_f = \omega_f^*$. Then, according to Whitehead,

$$h(f) = \int_{S^3} \omega_f^* \wedge \omega'_f \,,$$

and the required bound

$$h(f) \leq \text{const} \left(\text{Lip}(f) \right)^4$$

follows, since $\|\omega_f^*\| \leq \left(\text{Lip}(f) \right)^2 \|\omega\|$ and since one can always find a primitive ω' of ω^* satisfying

$$\sup_{x \in S^3} \|\omega'(x)\| \leq \text{const} \sup_{x \in S^3} \|\omega^*(x)\|,$$

as an elementary argument shows.

Sullivan's minimal models. These generalize the above construction and, assuming Y is simply connected, express *all* $\mathbb{R}$-valued invariants (i.e. homomorphisms) of the group $\pi_n(Y)$ as integrals of products of some pulled back differential forms on Y and their consecutive primitives. This leads to the bound

$$\|h^d\| \geq \text{const} \, d^\alpha$$

for every non-torsion element $h \in \pi_n(Y)$ where $\alpha = \alpha(Y, h) \neq 0$ is some *rational* number coming out of a computation with the minimal model.

Conjecture. Let Y be a compact simply connected Riemannian manifold (or a more general compact simply connected space with a "reasonable" metric). Then every non-torsion element $h \in \pi_n(Y)$ satisfies

$$\|h^d\| \sim d^\alpha$$

for the above mentioned rational number α provided by the minimal model.

What is unclear here is how to construct maps $f : S^n \to Y$ representing $[h^d]$ with $\text{Lip}(f) = O(d^\alpha)$. (Many such maps come via the Whitehead product and similar higher order products, but these seem to be unsufficient for our purpose.)

A special case of the above problem reads as follows. Let $h \in \pi_n(Y)$ be homologous to zero. Show that

$$\|h^d\| = O\left(d^{1/(n+1)} \right) \text{ for } d \to \infty.$$

In fact I do not even see how to get $\|h^d\| = o(d^{1/n})$ in this case.

Counting maps $\mathbf{X} \to \mathbf{Y}$. Denote by $\#(\lambda)$ the number of mutually non-homotopic maps $f : X \to Y$ with $\text{Lip}(f) \leq \lambda$. The above discussion tells us something about the asymptotics of $\#(\lambda)$ for $\lambda \to \infty$ in the case $X = S^n$ and in general, the minimal model method applies to all X and simply connected Y. Thus one can show

$$\#(\lambda) = O(\lambda^A), \quad \lambda \to \infty,$$

for some rational number A depending on the minimal models of X and Y, but one has a poor idea of how to generate sufficiently many homotopically distinct maps $f : X \to Y$ with small dilations in-so-far as the minimal model theory allows us to do it.

Controlled homotopy. Now, following the logic of our introductory remark on π_1 (Universe), we want to study the geometry of *homotopies* between maps $X \to Y$. For example, given a *contractible* map $f : X \to Y$ with $\mathrm{Lip}(f) \leq \lambda$, we want to find a homotopy f_t of $f = f_0$ to a point where each map $f_t : X \to Y$, $t \in [0, 1]$, has

$$\mathrm{Lip}(f_t) \leq \Lambda(\lambda)$$

for some "reasonable" function $\Lambda(\lambda)$. For example, if $X = S^n$, $Y = S^m$, where $n \neq m$, $2m - 1$, then the composed map

$$S^n \xrightarrow{\varphi} S^n \xrightarrow{f_0} S^m,$$

for some specific map $\varphi : S^n \to S^n$ of degree $k = k(m, n)$, is always contractible and we may ask our question for $f = \varphi \circ f_0$ where f_0 is an arbitrary map $S^n \to S^m$ with $\mathrm{Lip}(f_0) \leq \lambda_0$.

If Y is a non-simply connected space, then the (best possible) function $\Lambda(\lambda)$ may be essentially as complicated as any other recursive function, as follows from a recent (yet unpublished) work by Rips and Sapir. But for simply connected compact Riemannian manifolds one conjectures that

$$\Lambda(\lambda) \leq \mathrm{const}\, \lambda^p$$

for some p depending on the rational homotopy types of X and Y, where the minimal such $p = p(X, Y)$ is expected to be a rational number. (Actually, I have not worked out any example where $p > 1$.) What we know from the general principles of the homotopy theory is the bound

$$\Lambda(\lambda) \leq \exp\big(\exp \ldots (\exp(\lambda))\big)$$

where exp is iterated about $\dim X$ times. These exponents appear any time we appeal to the Serre fibration property as (uncareful) lifting of homotopies exponentially enlarges the dilation. However certain examples (such as S^{2n+1} fibred over $\mathbb{CP}^n$) suggest that one can get away with the polynomial enlargement.

Filling Riemannian manifolds. The (conjectural) bound $\Lambda(\lambda) = O(\lambda^p)$ has a (also conjectural) counterpart in the cobordism theory where we want to fill-in a closed Riemannian manifold V by W with a suitably defined size of W controlled by that of V. Here one may use the volumes of V and W for the size, provided one restricts to manifolds with appropriate bounds on their local geometries. Then one expects that every n-dimensional V bounds W, i.e. $\partial W = V$, such that

$$\mathrm{Vol}_{n+1} W \leq \mathrm{const}_n\, \mathrm{Vol}_n V\,,$$

where we assume that V is null-cobordant to start with. This conjecture is easy to prove for $n = 2$ while the only supporting evidence for $n \geq 3$ comes from our old result with Jeff Cheeger claiming that the η-invariant of a manifold V with bounded local geometry is bounded by $\mathrm{const}\,\mathrm{Vol}(V)$ for $\mathrm{const} = \mathrm{const}_n$ (bound on local geometry).

Morse landscape of the function Lip on the space Map(X → Y). This generalizes our previous perspective of counting homotopy classes of map $f : X \to Y$ with $\mathrm{Lip}(f) \leq \lambda$ as well as the controlled homotopy discussion. The Morse theoretic shape of the function Lip (e.g. the positions of its deep minima) is essentially independent of the specific metrics in X and Y and is determined by the homotopy types of these spaces. The question is how to effectively determine this shape for given X and Y.

Other dilation functions. Instead of Lip which measures the stretch of curves in X one can study some function

$$f \mapsto \mathrm{Lip}_j(f), \ f \in \mathrm{Map}(X \to Y),$$

measuring the stretch of j-dimensional submanifolds in X under f. In fact, one can look at several (all?) such functions Lip_j simultaneously and study the topology of the resulting map of the space $\mathrm{Map}(X \to Y)$ to some $\mathbb{R}^k$.

Noncompact spaces. Many interesting non-compact (e.g. infinite dimensional) spaces, (especially those of operator theoretic origin) come along with natural (classes of) metrics to which the above discussion applies. In fact, some information concerning the Lipschitz (and more general uniform) homotopy theoretic information can be derived from the classical isoperimetric inequality (via the measure concentration phenomenon of Levy–Milman) as we observed with Milman about 20 years ago. But this (measure theoretic) approach seems to be rather far removed from the above discussion which is more topological in origin.

References. See my forthcoming book "Metric structures for Riemannian and non-Riemannian spaces", Birkhäuser, 1998.

Acknowledgement. This paper was written on the basis of the notes of my talk prepared by Professor Hugo Rossi, to whom I express my gratitude.

INSTITUT DES HAUTES ÉTUDES SCIENTIFIQUES, DEPARTMENT OF MATHEMATICS, 91440 BURES-SUR-YVETTE, FRANCE
E-mail address: gromov@ihes.fr

Harmonic Analysis in Number Theory

Henryk Iwaniec

Preface

In this lecture I shall try to convey the idea of what modern Analytic Number Theory is about, what areas does it draw on and what does it give to other parts of mathematics? I am afraid there is no totally accurate image of the subject in the popular perception of number theory, for one often hears a question what, if anything, has analysis in common with arithmetic? Of course, these doubts do not matter as long as it is a "perfectly good and valid subject" to quote one of the most prominent arithmeticians. Thus, this lecture gives me an opportunity to explain that it is not wrong to employ analysis for studying properties of integers. To the contrary, these uncompromised combinations of methods are fruitful and delightful.

I don't think I will tell you everything that comprises analytic number theory. This is rather an informal talk on important topics in the theory which explore ideas from harmonic analysis. I realize no one alone can comprehend this huge territory. Fortunately, much of it is well popularized in numerous survey articles, not only in mathematics, but also in physics in the context of group representations. A new interaction between number theory and physics has recently appeared in arithmetic quantum chaos (see the stimulating article by P. Sarnak [**Sar**]). More importantly, the relationship between harmonic analysis and algebraic number theory has reached the state-of-the-art in the celebrated Langlands' program which was presented today. Thus, I confine my attention to these other, say classical, problems and methods. Though they require less abstract concepts, they are not always easy to describe precisely to a general audience in a one hour talk, so to make the exposition intelligible, sometimes I shall be deliberately vague. On many occasions I choose a simple result to illustrate an idea; for the strongest or more general results one needs to read particular publications which I selected in the bibliography. I have the courage to express my thoughts on various issues in the way they accompanied my studies; needless to say, these are not all original nor complete, and I would not expect everybody to agree with my views (it is refreshing for research when there are different opinions). I make no attempt either to keep the exposition in historical order or to give credits to all major inventions; such a task is beyond the scope of this talk.

1991 *Mathematics Subject Classification.* 11L07, 11M06.

1. How did it start?

To warm up let me make a few traditional comments. If you want to, you can observe a spirit of harmonic analysis whenever one discusses the generating power series

$$f(z) = \sum_{n=0}^{\infty} a_n z^n$$

which was a favorite language for Euler to speak of arithmetic properties of the sequence $\mathcal{A} = (a_n)$. Seriously, classical harmonic analysis began its service for number theory with expansion of a periodic function into Fourier series

$$f(z) = \sum_{n=-\infty}^{+\infty} \hat{f}(n)e(nx), \qquad e(z) = e^{2\pi i z} .$$

In modern exposition this amounts to considering the exponential functions $e(nx)$ as group characters (characters of the circle $\mathbf{R}/\mathbf{Z}$). Such a viewpoint takes us all the way back to Gauss, who invented the genus characters when studying the composition of binary quadratic forms. Then Dirichlet came up with characters on the multiplicative group $(\mathbf{Z}/q\mathbf{Z})^*$ of primitive residue classes $a \pmod q$ in connection with his proof that every such class has the same proportion of primes. Dirichlet is also credited for the introduction of series of the type

$$L_f(s) = \sum_{n=1}^{\infty} a_n n^{-s} .$$

This is the multiplicative analog of the power series. In the case of a character χ $(\mathrm{mod}\ q)$ we have a connection with prime numbers via the Euler product

$$L(s,\chi) = \sum_{1}^{\infty} \chi(n) n^{-s} = \prod_p (1 - \chi(p) p^{-s})^{-1} .$$

Therefore, thanks to the additive and multiplicative structures of integers, the idea of attaching a generating series to an arithmetic function has two realizations; today it flourishes beautifully in the garden of Hecke modular forms and in the "ghetto" of Maass wave forms.

I still need to mention the famous Riemann memoir in which the prime numbers are given dual companions—the complex zeros of the zeta-function

$$\zeta(s) = \sum_{1}^{\infty} n^{-s} = \prod_p (1 - p^{-s})^{-1} ,$$

$$\pi^{-s/2} \Gamma\left(\frac{s}{2}\right) \zeta(s) = \frac{1}{s(s-1)} \prod_{\zeta(\rho)=0} \left(1 - \frac{s}{\rho}\right) .$$

Here, according to the Riemann hypothesis, $\rho = 1/2 + i\gamma$. After Riemann, the exploration of functions of complex variables became a trademark of analytic number theory. Bear in mind that a representation of $-\zeta'(s)/\zeta(s)$ as the sum of simple fractions at the zeros of $\zeta(s)$ is a kind of spectral expansion.

For many arithmetic functions the corresponding generating series converges absolutely in a small domain, and it is an interesting question as to how far can the series be analytically continued? If the generating series extends beyond the range of absolute convergence, then this property usually manifests some sort of law for the distribution of the coefficients ("random" series cannot be continued). For example, the analytic continuation of the Artin L-function is intimately related to the reciprocity law in number fields (in the abelian case); the analytic continuation of the Hasse–Weil L-function is derived from the modularity of the corresponding elliptic curve (over $\mathbb{Q}$), and so on. Furthermore, a functional equation which connects values of L-functions at s and $1-s$, if it exists, is an analytic way of expressing some kind of law of symmetry which rules over the coefficients. Behind the functional equation one can find suitable summation operators which are self-adjoint.

2. The art of counting integers

On the perimeter of analytic number theory one sees plenty of asymptotic formulas and estimates for mean values of arithmetic functions. Certainly, a lot of these seem to be interesting in their own right to express properties which are not visible at individual places. Another motivation comes from the extraordinary practice of solving problems by way of counting the solutions. To put it very simply, if the estimate is positive, there must be a solution! Hence the question: how to count? This can be a very hard and long process. For example try to count primes of type $a^2 + b^4$. We have (a joint work with J. Friedlander [**FI**])

$$\#\{p \leq x : p = a^2 + b^4\} \sim cx^{3/4}(\log x)^{-1}, \qquad c = 1.1128\ldots,$$

but one needs more than ten fingers to establish this formula. The Grossencharacters of Hecke are employed and a lot of sifting as well. Notice that by this approach we have not built the primes in question in algebraic terms (for instance, we cannot parametrize the representations). Another illuminating example of roundabout approach to solving a problem is the Selberg proof of the existence of Maass cusp forms on the modular group by counting the discrete spectrum using the trace formula. Though the cusp forms appear in abundance, not a single one was ever constructed.

Now it should be clear that in order to be able to cultivate the fields of analytic number theory, as a prerequisite, one must acquire considerable skill for counting integral points inside various regular domains and on their boundaries.

One has no difficulty with counting integers in a segment; the number is approximately equal to the length of the segment and the error is bounded (one cannot do better). In a regular planar domain the number of lattice points approximates the area up to an error term whose order of magnitude is controlled by the length and shape of the boundary. The first approximation is obtained by the elementary method which uses packing with the unit square. For example, Gauss has established

$$(2.1) \qquad \sum_{n \leq x} r(n) = \pi x + O(x^{1/2})$$

where $r(n)$ denotes the number of ways that n can be written as the sum of two squares. Here πx is the area of the circle and $O(x^{1/2})$ is the bound for the circumference. Deeper approximations can be derived by Fourier expansion of the error term and estimating the relevant exponential sums.

A professional way of looking at the Gauss circle problem is to consider $X = \mathbf{R}^2$ as a homogeneous space acted on by the group of translations $G = \mathbf{R}^2$. The exponential functions $\varphi_{m_1, m_2}(x_1, x_2) = e(m_1 x_1 + m_2 x_2)$ are eigenfunctions of the

Laplace operator

$$D = \frac{\partial^2}{\partial x_1^2} + \frac{\partial^2}{\partial x_2^2}$$

with eigenvalues $\lambda_{m_1,m_2} = 4\pi(m_1^2 + m_2^2)$, and the well-known Fourier inversion is just the spectral resolution of a function satisfying proper decay conditions. When the function lives on the torus $\mathbf{R}^2/\mathbf{Z}^2$ (it is periodic) its spectral decomposition is just the classical Fourier expansion into $\varphi_{m_1,m_2}(x_1, x_2)$ with integral frequencies m_1, m_2. The trace formula on the torus is just the Poisson summation formula

$$\sum_{m_1}\sum_{m_2} f(m_1, m_2) = \sum_{n_1}\sum_{n_2} \hat{f}(n_1, n_2)$$

where $\hat{f}$ is the Fourier transform of f. In particular, for a radially symmetric function, say, $f(m_1, m_2) = k(m_1^2 + m_2^2)$, where k is smooth, compactly supported on $\mathbb{R}^+$, the Fourier transform is also radially symmetric (because the Laplacian commutes with rotations); precisely we have $\hat{f}(n_1, n_2) = h(n_1^2 + n_2^2)$ where h is the Hankel-type transform

$$h(v) = \pi \int_0^\infty k(u) J_0(2\pi\sqrt{uv}) \, du.$$

This gives us the Hardy–Landau–Voronoi summation formula

$$\sum_1^\infty r(m)k(m) = \pi \int_0^\infty k(u)du + \sum_1^\infty r(n)h(n).$$

Fiddling with the test functions $k(u)$, $h(v)$ one can derive the asymptotic formula (2.1) with improved error term $O(x^{1/3})$. Hardy showed this cannot hold with exponent $1/4$ whereas Huxley holds the world record $23/73 = 0.3150\ldots$. Huxley's arguments exploit extensively Gauss sums over finite fields and come down to the lattice point counting in some erratic domains of dimension ten (see [**Hux**]).

In view of the above connection the Gauss circle problem amounts to the Weyl law for the eigenvalues of the Laplace operator on the torus. It requires some imagination and considerable power to apply these ideas to other homogeneous spaces such as the modular surface $X = \mathrm{SL}_2(\mathbf{Z}) \setminus \mathbf{H}$ and the sphere $X = SO(3)/SO(2)$. In these spaces (Riemannian surfaces of curvature -1 and 1, respectively) the trace formula loses its self-duality; the geometric side is quite different than the spectral side (these are no longer sums over dual lattices). In the first case the Maass cusp forms (together with Eisenstein series) and in the latter the spherical harmonics (Legendre polynomials) replace the exponential functions on the torus.

The corresponding circle problem on the hyperbolic plane $\mathbf{H}$ is more sophisticated than that on $\mathbf{R}^2$. The aim is to estimate the number of points of the orbit $\{\gamma z : \gamma \in \Gamma\}$ which are at distance $\leq X$ from w (here z, w are chosen in $\mathbf{H}$ and Γ is a group acting discontinuously on $\mathbf{H}$). The natural method of packing with copies of the fundamental domain refuses to work for intrinsic reasons. Most of the area in a hyperbolic circle concentrates along the boundary (a lower segment of the circle) so it is comparable with the circumference, therefore the packing method yields an error term of order comparable with the main term—a meaningless result. For general domains this phenomenon is transparent in the isoperimetric inequality. It is the negative curvature which causes the trouble. We have already suggested that the Maass forms, say ϕ_j, are adequate harmonics to attack the problem. The analog of Poisson summation is the spectral decomposition

$$\sum_{\gamma \in \Gamma} k(u(\gamma z, w)) = \sum_j h(t_j)\varphi_j(z)\bar{\varphi}_j(w) + \text{ cont. spec. integrals}$$

where $u(z, w)$ is a suitable distance function on $\mathbf{H}$, $k(u)$ is a nice function on $\mathbf{R}^+$, and $h(t)$ is the Selberg and Harish–Chandra transform of k (cf. [**Hej**]). This is one of several gates through which non-abelian harmonic analysis (the spectral theorem for automorphic forms) sneaks into modern analytic number theory. The results are impressive. For example, if Γ is the modular group, one derives

$$|\{\gamma \in \Gamma : u(\gamma i, i) \le X\}| = 24X + O(X^{2/3}).$$

Let me translate this hyperbolic circle formula in terms of $r(n)$ to compare it with the Gauss circle problem. First of all, the hyperbolic case amounts to counting rational integers (a, b, c, d) on the determinant hypersurface $ad - bc = 1$ within the ball $a^2 + b^2 + c^2 + d^2 \le 4X + 2$ (we have $4u(i\gamma, i) + 2 = a^2 + b^2 + c^2 + d^2$ if $\gamma = \left(\begin{smallmatrix} a & b \\ c & d \end{smallmatrix} \right)$). After a linear change of variables the result reads as

$$\sum_{n \le x} r(n)r(n + 1) = 8x + O(x^{2/3}).$$

Such a strong asymptotic cannot be derived by repeated applications of classical Poisson summation. Well, not literally speaking, because one can hire additive characters to reduce the problem to sums of Kloosterman sums, but to estimate the latter one needs again the spectral theorem for automorphic forms (by Weil bound for individual Kloosterman sums one could derive a weaker result with exponent $5/6$ in place of $2/3$). We have chosen to follow a more direct path (see Chapter 12 of [**I4**]).

The error term $O(x^{2/3})$ in the above result, or in general hyperbolic circle problems, has never been improved even slightly for any Fuchsian group. Recently, R. Phillips and Z. Rudnick [**PR**] gave insightful analysis of the matter from a statistical point of view, while F. Chamizo [**Cha**] established sharp estimates for the L_2-norm of the error term. Both works suggest that the true error term should be $O(x^{1/2+\varepsilon})$.

Even more impressive results come out from harmonic analysis on the sphere $\mathbf{S}^2 = SO(3)/SO(2)$. W. Duke [**Duk**] proved (using estimates for Fourier coefficients of half-integral weight cusp forms, see [**I3**]) that the integral points (m_1, m_2, m_2) on the sphere $m_1^2 + m_2^2 + m_3^2 = n$ are equidistributed (with respect to the Haar measure on the surface of the sphere) as n grows over numbers satisfying some local conditions (the total number of points on the sphere is essentially the class number of the imaginary quadratic field $\mathbf{Q}\left(\sqrt{-n}\right)$). In this case the corresponding Weyl sums from the theory of equidistribution are essentially the n-th coefficient of metaplectic cusp forms, and one needs to beat the convexity bound for these coefficients. The latter is a problem which involves another idea of harmonic analysis, to be discussed in Section 5.

Without doubt the harmonic analysis on higher rank groups will also make its permanent home in analytic number theory in the near future. There are already interesting developments, in a large measure due to the influence of P. Sarnak (see his stimulating ICM Kyoto lecture).

By the way, when we count points in the hyperbolic circle, a mixture of two problems emerges; that of the diophantine equation $ad - bc = 1$ and that of the lattice points in the ball $a^2 + b^2 + c^2 + c^2 \leq T^2$. Far more general problems are considered in what is called diophantine analysis. Let V be an affine variety given by integral polynomials and let $V(\mathbf{Z})$ denote the set of $\mathbf{Z}$-points of V. One would like to know the asymptotic behavior of

$$N(T, V) = \left| \{ m \in V(\mathbf{Z}) : \|m\| \leq T \} \right|.$$

If $V(\mathbf{R})$ is a symmetric space (which is defined via action of a linear algebraic group G), then some satisfactory results have been established by employing harmonic analysis on $G(\mathbf{Z}) \setminus G(\mathbf{R})$; for example see [**DRS**]. However, in complete generality the problem is hopeless. The old circle method of Hardy–Littlewood is available but it works only if there are sufficiently many points on $V(\mathbf{Z})$ (cf. [**Bir**]).

I should also mention the recent works of A. Eskin, G. A. Margulis, and S. Mozes [**EMM**] about small values of indefinite ternary quadratic forms (quantitative versions and effectivizations of the Oppenheim conjecture). They proceed with a different way of counting lattice points (the points of special trajectories of $\mathrm{SL}_3(\mathbf{R})/\mathrm{SL}_3(\mathbf{Z})$), while still applying analysis on non-commutative groups they use ergodic-theoretic rather than spectral-theoretic ideas.

3. In quest of arithmetic harmonics

I have mentioned that the circle method of Hardy–Littlewood is capable of producing an asymptotic formula for the number of solutions to some diophantine equations. The mechanisms of the method are explained from this perspective in [**Sch**]. Recall, however, that the circle method itself was invented in the seminal paper of Hardy–Ramanujan [**HR**] on the partition function. What more that needs to be said is the method (not a refinement) due to Kloosterman [**Klo**]. Kloosterman works lie far deeper than what he is recognized for. I refer to the ICM Stockholm address by Yu. V. Linnik [**Lin1**] for an accurate account. In this section I indulge myself in speculations of a general nature which have to do with the possible extensions of Kloosterman's ideas. I suspect you will find my speculations to be inconclusive. This is because the research is still in its initial stage so I will not insist on introducing definitive concepts.

Many problems in number theory lead to the question of whether one set of integers meets another. Thus it is the question of solvability of the equation $m = n$ with m, n running independently over our chosen sets. Generalizing slightly, we wish to estimate the sum

$$(3.1) \qquad\qquad S(f, g) = \sum_m f(m)\overline{g(m)}$$

where $f, g : \mathbf{N} \to \mathbf{C}$ are suitable arithmetic functions. Our knowledge about f and g in practice is programmed in some kind of generating series. A generating series for f is obtained by twisting with a distinct arithmetic function $\chi : \mathbf{N} \to \mathbf{C}$

$$(3.2) \qquad\qquad S(f, \chi) = \sum_m f(m)\chi(m).$$

We call χ a "harmonic"; it is an oscillatory function such as an additive or multiplicative character, the multiplicative function m^{it}, Hecke eigenvalues, and so on (there are plenty of possibilities that derive from two dimensional representations).

Our choice of harmonics for the problem at hand is usually self-suggesting. An adequate χ must "cooperate" with arithmetic characteristics of f so that the twisted series $S(f, \chi)$ satisfies a suitable transformation rule (it could be only an approximate equation with small error term), most likely of an involutory type such as the functional equation for automorphic L-functions or the approximate Poisson's formula for exponential sums (called process B of van der Corput, see [**GK**]). There is a limit to what can be done with a single transformation rule; it is determined by the uncertainty principles of harmonic analysis. A typical example of this principle is the relation between essential domains of support of a function and its Fourier transform; when one shrinks to a point, the other expands over the whole space.

To reduce the effect of the uncertainty one employs a variety of harmonics; these must be independent in a sense of orthogonality, or asymptotic orthogonality, in the ambient space. For example, the two harmonics $\chi(n) = n^{it}$ and $\chi_1(n) = n^{i(t-1)}$ are too close to each other to project distinctly, in spite of being linearly independent as functions on $\mathbf{N}$. Some harmonics have more power than others; it is measured by a quantity which is reminiscent of the conductor of a Dirichlet character. Let us not dwell on precise definitions because what is truly behind the notion of conductor can be explained descriptively as follows. We begin by testing $S(f, \chi)$ for a nice smooth function f supported on a dyadic interval of length $M \geq 1$ such that $f^{(j)} \ll M^{-j}$. If M is not too large, by the transformation rule, the normalized sum $S(f, \chi)M^{-1/2}$ is changed into $\varepsilon_\chi S(f', \chi')N^{-1/2}$ where $|\varepsilon_\chi| = 1$ and f' is supported on a dyadic interval of length $N \geq 1$ up to a small remainder term. When $Q = MN$ is more or less constant, we call this quantity the conductor of χ. Therefore, a smooth sum of $\chi(m)$ which is longer than the square root of the conductor turns into a shorter sum and vice-versa. The smaller the conductor the more powerful is the harmonic. There is some consistency in the behavior of twisted sums of arithmetic functions in particular families which are chosen for the harmonics under considerations. Usually, if χ has conductor Q then for an arithmetic function f in the family, the connection between $S(f, \chi)M^{-1/2}$ and $\varepsilon_\chi(f)S(f', \chi')N^{-1/2}$ holds with $MN = Q^d$ where d is a positive integer (this connection is tied at $M = N = Q^{d/2}$). For example, take the sequence of Hecke eigenvalues or the sequence of values of a binary quadratic form truncated smoothly to the interval $M < m < 2M$. In these cases $d = 2$. In general, $d = d_f$ represents the degree of f. The complex numbers $\varepsilon_\chi(f)$ depend very strongly on the individual harmonic χ and rather simply on f; actually quite often they depend only on $d = \deg f$, precisely $\varepsilon_\chi(f) = \zeta \varepsilon_\chi^d$ where ζ is a constant. For $\chi \pmod{q}$ a primitive Dirichlet character, ε_χ is the normalized Gauss sum.

Now we return to the original sum $S(f, g)$; we write this as

$$S(f, g) = \sum_m \sum_n \delta(m, n) f(m) g(n)$$

where $\delta(m, n)$ is the diagonal symbol of Kronecker. With a variety of orthogonal harmonics which are adequate for both f and g we can pick up the diagonal $m = n$ exactly, or amplify its contribution so that the off-diagonal terms are negligible. Paraphrasing, we seek a good approximation for the diagonal symbol of type

$$\delta(m, n) \sim \sum_\chi c_\chi \chi(m) \bar{\chi}(n)$$

getting

$$S(f,g) \sim \sum_{\chi} c_{\chi} S(f,\chi)\bar{S}(g,\chi)\,.$$

Applying the transformation rules to the sums $S(f,\chi)$ and $S(g,\chi)$, we get

$$S(f,g) \sim M^2 Q^{-\frac{1}{2}(d_f+d_g)} \sum_m \sum_n f'(m)\overline{g'(n)}\Delta(m,n)$$

where $\Delta(m,n)$ is a kind of Fourier transform in the space of harmonics

$$\Delta(m,n) = \sum_{\chi} c_{\chi}\varepsilon_{\chi}(f)\overline{\varepsilon_{\chi}}(g)\chi'(m)\overline{\chi'}(n)\,.$$

If the space of harmonics is fairly complete in some spectral sense, there is a dual side of $\Delta(m,n)$, say

$$\Delta(m,n) \sim \sum_{\xi} \xi(m,n)\,,$$

where the $\xi(m,n)$ are rather different from the $\chi'(m)\overline{\chi'}(n)$. Moreover, if the space of harmonics is quite large, we receive on the dual side only a few terms $\xi(m,n)$ which are also simpler to deal with than many of the original harmonics (so one profits reasonably by going to the dual side of $\Delta(m,n)$ not only from reducing the quantity but also from changing the structure of harmonics). Hence we arrive at the approximation

$$S(f,g) \sim M^2 Q^{-\frac{1}{2}(d_f+d_g)} \sum_{\xi} \sum_m \sum_n f'(m)\overline{g'(m)}\xi(m,n)\,.$$

From this point on it is hard to continue in a unified fashion. In order to proceed further we need to specialize. Before giving examples I mention briefly that these ideas were used in [**DI1**] to re-establish meromorphic continuation of the symmetric cube L-function (attached to a Hecke cusp form on the modular group) up to the critical line (recently H. Kim and F. Shahidi [**KiSh**] proved that the symmetric cube L-functions are entire). Another case is considered in [**DI3**]. In this application χ runs over Dirichlet characters, $f(m)$ and $g(m)$ are the coefficients of the Shimura symmetric-square L-function (these have degree 3), and the result gives an estimate for the fourth power-moment of the Hecke eigenvalues.

Now it is time to illustrate the above abstract scheme by concrete realizations. I will show two cases; the first one comes from the Kloosterman circle method, the second one from the Petersson formula in modular forms. In the circle method we employ the integral (Vinogradov refinement)

$$\delta(m,n) = \int_0^1 e(\alpha(m-n))d\alpha$$

and obtain

$$S(f,g) = \int_0^1 S_f(\alpha)S_g(\alpha)d\alpha$$

where $S_f(\alpha)$ denotes the corresponding exponential sum. Following Kloosterman each α is approximated by a rational number a/c with $c \leq C$ and $(a,c) = 1$ so that $S_f(\alpha)$ is not much different from $S_f(a/c)$ and the latter is evaluated by an appeal to special features of f. Think of the points a/c as cusps for the modular group, I suppose Ramanujan would not object. These fractions are not evenly distributed modulo 1, which causes certain levelling problems (one has to take care of overlapping arcs). This problem can be resolved neatly by the aid of the following expansion

$$(3.3) \qquad \delta(m,n) = \int_0^1 \sum_{\substack{0<c\leq C<d\leq c+C \\ (c,d)=1}} 2(dc)^{-1} \cos\left(2\pi(m-n)\left(\frac{a}{c}-\frac{\alpha}{cd}\right)\right) d\alpha$$

where $ad \equiv 1(\bmod\, c)$. Here C is at our disposal, with a good choice being $C = 2\sqrt{N}$ if $|m|, |n| \leq N$ because it makes the integration manageable. The formula (3.3) gives us a representation of the diagonal symbol $\delta(m,n)$ in terms of additive characters $e((m-n)a/c)$ of relatively small modulus c in addition to a mildly oscillating factor $e(\alpha(m-n)/cd)$. This small perturbation can be eliminated by a standard Fourier method. Moreover, the range for d is quite stable, so it also can be detected by a standard Fourier method at small price. This brings us to the Kloosterman sums

$$S(h,k;c) = \sum_{ad\equiv 1\ (\bmod\, c)} e\left(\frac{ah+dk}{c}\right)$$

where $h = m - n$ and k is a small integer. Let me emphasize that the appearance of Kloosterman sums is the aftermath of the levelling operation; it should not be confused with the exponential sums of similar nature which emerge from $\Delta(m,n)$ if f and g are quadratic forms as in Kloosterman's original work.

One can easily detect the diagonal by the simpler identity

$$\delta(m,n) = \frac{1}{q} \sum_{a(\bmod\, q)} e\left((m-n)\frac{a}{q}\right) = \frac{1}{q}\sum_{c|q} R_c(m-n)$$

where $R_c(h) = S(h,0;c)$ is the Ramanujan sum; however this identity requires $q \geq |h|$, so it contains characters to extremely large moduli and applications are limited very much. Another identity has been recently proposed in place of (3.3), namely

$$(3.4) \qquad \delta(m,n) = \sum_c R_c(m-n)\Delta_c(m-n)$$

where

$$\Delta_c(u) = \sum_{q\equiv 0\ (\bmod\, c)} q^{-1}\left(\omega(q) - \omega(|u|q^{-1})\right).$$

Here $\omega(t)$ is any smooth and compactly supported function on $\mathbf{R}^+$ with $\sum\omega(n) = 1$ (the normalization condition). We show that $\Delta_c(u)$ approximates the Dirac distribution, i.e.

$$\int_{-\infty}^{+\infty} f(u)\Delta_c(u)du \sim f(0)$$

on a certain class of smooth test functions (see [**DFI2**]). Choosing $\omega(t)$ supported on the dyadic interval $[\sqrt{N}, 2\sqrt{N}]$ one gets an expansion for $\delta(m,n)$ in terms of additive characters $e((m-n)d/c)$ to modulus $c < 2\sqrt{N}$. Both identities (3.3) and (3.4) have the same potential for applications. Perhaps a slight advantage of the latter is that it is free of Kloosterman sums.

When talking about the circle method it is appropriate to point out some character issues. As far as the harmonic analysis on the group $(\mathbf{Z}/q\mathbf{Z})^*$ is concerned everything seems to be known—not at all! Let me ask the question: given a character $\chi(\mathrm{mod}\,q)$, how far has one to search for the first $a > 0$ such that $\chi(a) \neq 1$? Probably there exists such $a \ll (\log q)^2$, but we cannot prove this bound without using the Riemann hypothesis.

Let me say briefly what has been done in the absence of the Riemann hypothesis. Three decades ago a new direction began in analytic number theory by producing the ℓ_2-estimates for character sums which turned out to be extremely powerful so as to eliminate the need for the Riemann hypothesis in numerous applications. The idea originated from Linnik's short paper [**Lin2**] on the large-sieve method; thus they called the results "large-sieve inequalities". Linnik showed that the least $a > 1$ with $\chi_p(a) = -1$ satisfies $a \ll p^\varepsilon$ for any $\varepsilon > 0$ with a very few exceptions. Major contributions after Linnik were made by E. Bombieri and H. L. Montgomery (see his comprehensive article [**Mon**]). One (of several diverse) analytic principles of the large sieve uses the duality in a finite dimensional Hilbert space (it asserts that an operator and its adjoint have the same norm). I will comment on a few fundamental results. First is the inequality of Bombieri [**Bom**] for linear forms in primitive additive characters

$$\sum_{q \leq Q} {\sum_{a(\mathrm{mod}\,q)}}^* \left| \sum_{n \leq N} a_n e\left(n\frac{a}{q}\right) \right|^2 \ll (Q^2 + N) \sum_{n \leq N} |a_n|^2\,.$$

Hence, using Gauss sums, he derived the same bound for the multiplicative primitive characters $\chi(n)$ in place of $e(na/q)$. R. Heath-Brown [**H-B**] considered only real characters (a much harder problem), showing that

$$\sum_{m \leq M}^\flat \left| \sum_{n \leq N}^\flat a_n \left(\frac{m}{n}\right) \right|^2 \ll (MN)^\varepsilon (M + N) \sum_{n \leq N}^\flat |a_n|^2$$

where $\sum^\flat$ restricts the summation to squarefree numbers.

So far I spoke only about characters on the group of residue classes. Starting about 1980 analytic number territory expanded into the theory of automorphic forms. The Fourier coefficients of cusp forms are given equal status with characters in the space of arithmetic harmonics. We have, among other types of large sieve inequalities the following result

$$\sum_{k \leq K} \sum_{f \in S_k(q)} \left| \sum_{n \leq N} a_n \lambda_f(n) \right|^2 \ll (qKN)^\varepsilon (qK^2 + N) \sum_{n \leq N} |a_n|^2$$

where $S_k(q)$ denotes the family of primitive cusp forms of level q and weight $k \geq 2$, and $\lambda_f(n)$ is the Hecke eigenvalue.

Let $\{f\}$ be an orthonormal basis of the whole space $S_k(q)$ of cusp forms on $\Gamma_0(q)$ of level q and weight $k > 2$ equipped with Petersson's inner product. Let $\hat{f}(n)$ denote the n-th Fourier coefficient of f we scale it down to

$$\psi_f(n) = \Gamma(k-1)^{1/2}(4\pi n)^{\frac{1-k}{2}} \hat{f}(n)\,.$$

If f is an eigenform of all the Hecke operators, then $\psi_f(n) = \lambda_f(n)\psi_f(1)$. For the purpose of analytic number theory I would like to think of the $\psi_f(n)$ as being random variables. In the terminology previously developed, the $\psi_f(n)$ are harmonics of conductor q, very similar to the Dirichlet characters $\chi(\mathrm{mod}\,q)$ in spite of being

GL$_2$ objects. This observation has profound consequences which I will bring up later.

The new harmonics $\psi_f(n)$ with $f \in S_k(q)$ are capable of detecting the diagonal symbol $\delta(m, n)$ more efficiently than their cousins $\chi(\mathrm{mod}\, q)$ in special cases. On the one hand we form the spectral sum

$$\Delta(m, n) = \sum_f \psi_f(m)\overline{\psi_f}(n)\,;$$

on the other hand we have the sum of Kloosterman sums

$$\sigma(m, n) = 2\pi i^k \sum_{c \equiv 0 \,(\mathrm{mod}\, q)} c^{-1} S(m, n; c) J_{k-1}\left(\frac{4\pi\sqrt{mn}}{c}\right)$$

where $J_{k-1}(x)$ is the Bessel function (it is itself a continuous analog of a Kloosterman sum). Both are related by

(3.5) $$\Delta(m, n) = \delta(m, n) + \sigma(m, n)\,.$$

This exact formula is a synthesis of various results by H, Petersson, R. Rankin and A. Selberg.

For simplicity we have made our analysis more restrictive than it should be. Indeed, it is necessary to engage a complete set of automorphic forms in the space $L^2(\Gamma_0(q) \backslash \mathbf{H})$ in order to utilize the whole spectrum of the Laplace operator. There are formulas analogous to (3.3) in the space $L^2(\Gamma_0(q) \backslash \mathbf{H})$ developed by R. Bruggeman [**Bru**] and N. Kuznetsov [**Kuz**]. These are useful in two directions: to study the Fourier coefficients of automorphic forms or sums of Kloosterman sums.

4. The determinant equation, Kloosterman sums and closed geodesics

The Kloosterman sums have been used to solve difficult problems in analytic number theory since the time of their creation. I will not show how this is done because any interesting example involves ideas which go far beyond the scope of this talk. Briefly speaking, the problems are reduced (with some ingenuity) to the determinant equation

(4.1) $$\det \begin{pmatrix} a & b \\ c & d \end{pmatrix} = n\,,$$

and the Kloosterman sums are instrumental for counting the solutions. One would like to have, for a fixed integer $n \neq 0$, a good asymptotic formula for the number of solutions as the entries vary over certain sequences of integers. In this generality, the problem is intractable (it would have striking implications, for example to the twin prime problem). If we let the entries run over fixed arithmetic progressions, this is a problem from the spectral theory of GL$_2$ automorphic forms. Not only does the Laplacian play a role, but the Hecke operators do as well.

One can also deal with the determinant equation exclusively within the framework of Kloosterman sums using the best possible estimates for individual sums derived from the Riemann hypothesis for curves by A. Weil. The results are good, but the spectral theory yields better things. But not immediately! Here, and in many similar problems, there is a question of how small is the first positive eigenvalue λ_1 of the Laplacian on $L^2(\Gamma \backslash \mathbf{H})$ for congruence groups? The eigenvalues $0 < \lambda_j < 1/4$, if they are there, damage the asymptotic formula for a sum of Kloosterman sums similarly to what the exceptional zero of the Dirichlet L-function does

to the explicit formula for primes in arithmetic progressions. A comparable defect
appears in the asymptotic formula for the length of closed geodesics on $\Gamma \backslash \mathbf{H}$. One
shouldn't underestimate the arithmetical nature of the problem. Selberg showed
that $\lambda_1 \geq 3/16$ by an appeal to Weil's bound for Kloosterman sums, and he conjectured that $\lambda_1 \geq 1/4$. It is not surprising that Selberg's bound is not sufficient
to produce improvements in estimates for sums of Kloosterman sums; after all it
uses the Weil bound. However, recently W. Luo, Z. Rudnick and P. Sarnak [**LRS**]
succeeded in showing that $\lambda_1 \geq 171/784$. A simpler proof of a slightly weaker result $\lambda_1 \geq 10/49$ was given in [**I2**]. This is a triumph of diverse ideas, among them
implicitly the harmonic analysis on GL_3 and exponential sums over algebraic varieties (Deligne's estimate for hyper-Kloosterman sums; a similar combination of
arguments is present in much earlier work [**DI2**]). From the above lower bound for
the Laplace eigenvalue, W. Luo and P. Sarnak [**LS**] deduced that

$$(4.2) \qquad \sum_{\substack{c \leq x \\ c \equiv 0 \ (\mathrm{mod}\ q)}} c^{-1} S(m,n;c) \ll x^{5/14},$$

and

$$(4.3) \qquad \sum_{NP \leq x} \log NP = x + O(x^{7/10})$$

where P runs over the hyperbolic conjugacy classes in $\Gamma_0(q) \backslash \mathbb{H}$ and $\log NP$ is
the length of the corresponding closed geodesic. In the closed geodesic theorem the
error term $O(x^{7/10})$ is better than $O(x^{3/4})$ which is easily obtained by employing the
Riemann hypothesis for the Selberg zeta-function. The above improvement requires
some regularity in the spacing of zeros on the critical line which is established by
appealing to the Rankin–Selberg L-functions. A corresponding improvement for the
sum over rational primes is not possible because there is not sufficient regularity in
the spacing of zeros of the Riemann zeta-function.

We offer similar observations about (4.2). First notice that by summing with
absolute values one gets the bound $x^{1/2}$ in place of $x^{5/14}$. Thus, this improved
estimate shows there is a cancellation in sums of Kloosterman sums which is due
to the variation of sign of $S(m,n;c)$ as c ranges over the multiples of q (one can get
the same estimate for sums over c in any fixed residue class modulo q).

When we are talking about the variation in sign of Kloosterman sums, it is an
opportunity to highlight another direction undertaken by N. Katz [**Kat**]. Writing
$S(a,1;p) = 2\sqrt{p} \cos \theta_p(a)$ with $0 < \theta_p(a) < \pi$, Katz proved that the angles $\theta_p(a)$
are equidistributed with respect to the Sato–Tate measure $\frac{2}{\pi}(\sin \theta)^2 d\theta$ as a varies
modulo p and p tends to infinity. The same distribution law is expected to hold for
any fixed $a \neq 0$ as p varies over primes. The most advanced results in the latter
direction have been established by P. Michel [**Mic**].

While the harmonic analysis on GL_3 helps to improve working conditions in
the space of GL_2 forms by providing strong estimates for eigenvalues (at finite and
infinite places), there is not yet a direct route from GL_3 to the classical problems.
For example, we would like to have an adequate spectral theory of the equation

$$(4.4) \qquad x_1 x_2 x_3 - x_4 x_5 x_6 = n$$

(this is not given by the determinant in GL_3), but so far it has not been established.
Thus we had to treat (4.4) as the determinant equation for the congruence group
$\Gamma(x_1, x_4)$ pretending x_1 and x_4 are fixed integers. This approach to the problem

requires immense care to deal with the uniformity in terms of the level of the group, and the latter brings with it a further difficulty with small eigenvalues. A few eigenvalues which do not satisfy Selberg's conjecture do not play a role but with a large number of these one should question the sense of using spectral theory in the first place. Fortunately, for the Selberg conjecture we have a sufficiently good substitute which is the following density estimate

$$\# \left\{ j; \lambda_j < \tfrac{1}{4} - r^2 \right\} \ll V^{1-4r+\varepsilon}$$

where $V = \text{Vol}(\Gamma \setminus \mathbf{H})$ is the volume (see [**I1**]). Although this approach is not perfect, it produces significant improvements (unthinkable to obtain by conventional methods).

I would like to finish this section by describing two results, proofs of which are a synthesis of the above methods and many other methods from classical areas of analytic number theory (exponential sums and sieve methods). Let $\pi(x; q, a)$ denote the number of primes $p \leq x$ with $p \equiv a(\bmod q)$. We then have

$$\sum_{\substack{q < Q \\ (q,a)=1}} \lambda(q) \left(\pi(x; q, a) - \frac{\pi(x)}{\varphi(q)} \right) \ll x(\log x)^{-A}$$

with $Q = x^{4/7-\varepsilon}$ for any ε, $A > 0$, where the implied constant depends on ε, a, A. Here λ is any bounded function with certain properties which are not restrictive for basic applications (especially for applications of sieve methods); it is called a well-factorable function after these properties. This result was established by the joint efforts of E. Bombieri, E. Fouvry, J. Friedlander and myself. It shows that primes are evenly distributed over primitive residue classes on average over very large moduli. Our range of moduli stretches beyond the capability of the Riemann hypothesis; the latter stops working at $Q = x^{1/2}$. All these works (see [**BFI**] and [**Fou**]) are empowered by the spectral theory of automorphic forms indirectly when we make an appeal to Kloosterman sums and the analysis of the determinant equation (4.1) with very intricate entries.

A new treatment of the determinant equation (4.1) has been recently proposed in [**DFI4**] using the method of amplification, which I shall describe in the next section, but rather in a context which is suitable for automorphic L-functions.

Our second result, which is based heavily on spectral theory of automorphic forms, concerns roots of the congruence $f(\nu) \equiv 0 \pmod{p}$, where $f(x) = ax^2 + bx + c$ is a quadratic, irreducible polynomial with integer coefficients. If p is sufficiently large there are two or no solutions $\nu(\bmod p)$ according to whether $\Delta = b^2 - 4ac$ is or is not a quadratic residue modulo p. We (see [**DFI1**] and [**Tot**]) showed that the fractions ν/p are asymptotically equidistributed modulo one; precisely, for any smooth function $\psi : \mathbb{R}/\mathbb{Z} \to \mathbb{C}$ we have

$$\lim_{x \to \infty} \frac{1}{\pi(x)} \sum_{p \leq x} \sum_{f(\nu) \equiv 0 \ (\bmod p)} \psi\left(\frac{\nu}{p}\right) = \int_0^1 \psi(\alpha) \, d\alpha \, .$$

5. The amplification method

Here I give some idea what the method of amplification is about. It shows, among other things, more clearly than ever before how powerful the argument of positivity is in conjunction with harmonic analysis. To save time, we continue the presentation in the same context as in Section 3, which is designed for applications to automorphic L-functions. We return to the sum of an arithmetic function twisted by special harmonics

$$(5.1) \qquad S(f,\chi) = \sum_{m \sim M} f(m)\chi(m) \,.$$

We normalize so that f and χ are bounded on average. Summing with absolute values gives us the trivial bound $S(f,\chi) \ll M^{1+\varepsilon}$. There is a general philosophy which implies that such sums should be bounded by the square root of the number of terms due to the cancellation; precisely that $S(f,\chi) \ll M^{1/2+\varepsilon}$, subject to some mild conditions on f and χ. This philosophy generalizes the Lindelöf hypothesis for L-functions, and it is usually a consequence of the Riemann hypothesis if applicable.

Applying the reflection rule $S(f,\chi)M^{-1/2} \rightthreetimes S(f',\chi')N^{-1/2}$ where $MN = Q^d$, we may assume we are on the shorter side of this equation, i.e. that $M \ll Q^{d/2}$. Then the trivial bound yields

$$(5.2) \qquad S(f,\chi) \ll Q^{\frac{d}{2}+\varepsilon} \,.$$

This is called a convexity bound because when $S(f,\chi)$ is a partial sum of an L-function, such a result is derived by a relevant functional equation and the convexity theorem of Phragmen–Lindelöf. In applications (to real problems, at any rate) of the convexity bound one just misses the target; however even a small improvement would suffice. On the other hand, we rarely need a bound as strong as that provided by the Riemann hypothesis. Many problems can be given a full solution with the aid of any subconvexity bound. For example, the complete solution of the Linnik problem about equidistribution of points on a sphere requires improving only the convexity bound for automorphic L-functions twisted by quadratic character (via Waldspurger theorem and the Shimura correspondence). To say "only" is an understatement; the convexity bound, though trivial to reach, is a difficult barrier to cross. H. Weyl [**Wey**] succeeded with the Riemann zeta-function, and D. Burgess [**Bur**] with the Dirichlet L-functions. Their methods are different and rather special.

Now we describe very briefly a new method for breaking the convexity bounds which applies widely to the L-functions of degree 2 (see the survey article by J. Friedlander [**Fri**]). First we wish to majorize the sum $S(f,\chi)$ by something manageable which does not contain the harmonic χ since all its properties that are available to us are already exploited. For this reason we first adopt for χ a family of companions, say $\mathcal{H}$, and try to estimate the whole sum

$$(5.3) \qquad \mathcal{L} = \sum_{\chi \in \mathcal{H}} |S(f,\chi)|^2 \,.$$

Squaring out and changing the order of summation we get

$$(5.4) \qquad \mathcal{L} = \sum_{m_1} \sum_{m_2} f(m_1)\bar{f}(m_2)\Delta(m_1,m_2)$$

where

$$\Delta(m_1,m_2) = \sum_{\chi \in \mathcal{H}} \chi(m_1)\bar{\chi}(m_2) \,.$$

Now, assuming that $\mathcal{H}$ is a complete family, or almost complete in some asymptotic sense, there is a dual side to $\Delta(m_1,m_2)$ of type

$$\Delta(m_1,m_2) = \delta(m_1,m_2)H + \sigma(m_1,m_2)$$

where $H = |\mathcal{H}|$ is the number of family members and $\sigma(m_1, m_2)$ is the contribution from the off-diagonal terms. This contribution is expected to be small because we require the almost orthogonality of our harmonics. Notice that these intractable harmonics are absent on the dual side. The summation of the off-diagonal terms can be easier or more difficult to carry out than the original sum (5.1) depending on how large the family of harmonics is. According to the uncertainty principle the larger the family of companions is the simpler the dual side should be. Estimating $\sigma(m_1, m_2)$ trivially we obtain the following mean-value formula

$$(5.5) \qquad \sum_{\chi \in \mathcal{H}} |S(f, \chi)|^2 = \{H + O(M)\} \sum_m |f(m)|^2 .$$

We are often able to reduce the average contribution of the off-diagonal term $f(m_1)\bar{f}(m_2)\sigma(m_1, m_2)$ due to cancellation. This is roughly equivalent to the fact that "something or other is equidistributed" (for example the CM-points, as in [**Duk**]). We obtain

$$(5.6) \qquad \sum_{\chi \in \mathcal{H}} |S(f, \chi)|^2 = \{H + O(M^{1-\delta})\} \sum_m |f(m)|^2$$

where δ is a positive constant.

In practice the requirement of completeness forces the family to be at least as large as the conductor, $H = |\mathcal{H}| \asymp Q$. Notice that in such cases the mean-value theorem is best used if $d = \deg f = 2$ because $M \asymp Q^{d/2}$. Therefore we restrict further presentation to arithmetic functions of degree two, so $H \asymp M \asymp Q$. An interesting case is the Artin L-function attached to a two-dimensional Galois representation, in particular to a dihedral one. In this case a complete family of harmonics is given by the class characters for an imaginary quadratic field $K = \mathbb{Q}(\sqrt{D})$; its cardinality is equal to the class number $H = |\mathcal{H}| = h(D)$, which satisfies $|D|^{\frac{1}{2} - \varepsilon} \ll h(D) \ll |D|^{\frac{1}{2}} \log |D|$. One may also choose a larger family of harmonics by including other representations (tetrahedral, octahedral, icosahedral) and Maass forms of weight one with the Laplace eigenvalue near $1/4$. The dual side in the first choice is essentially a character sum over the CM-points, whereas for the enlarged family it is a sum of Kloosterman sums. Clearly, the nature of the latter is more friendly than that of the former. However, one should have some reservation for enlarging too much the family of companions because the individual members must have relatively significant contribution.

Ignoring all but one term on the left side of (5.6) one just recovers the trivial bound for $S(f, \chi)$. In general this is the best result one can hope for. To improve this bound we must reduce the contribution of the diagonal on the dual side. We accomplish this by introducing a twist by $\chi(\ell)$, getting the following bound (because of sign change of characters):

$$(5.7) \qquad \sum_{\chi \in \mathcal{H}} \chi(\ell)|S(f, \chi)|^2 \ll \{H\ell^{-1/2} + M^{1-\delta}\ell^A\} \sum_m |f(m)|^2$$

for some constants $\delta > 0$, $A > 0$. This result exploits quite effectively the multiplicativity of harmonics, in particular the theory of Hecke operators. A special attraction of these harmonics is that one can build $\chi(\ell)$ into $\chi(m)$ inside of $S(f, \chi)$ so that one is still able to use the spectral decomposition for sums over $\mathcal{H}$. If ℓ is sufficiently small, then (5.7) is better than (5.6). Unfortunately, the positivity on the left side is lost. To restore the positivity we average over ℓ, and we derive from (5.7) that

$$\sum_{\chi \in \mathcal{H}} |S(\lambda, \chi)|^2 \, |S(f, \chi)|^2 \ll (Q + Q^{1-\delta} L^A) \Big(\sum_\ell |\lambda(\ell)|^2\Big) \Big(\sum_m |f(m)|^2\Big)$$

where $S(\lambda, \chi) = \sum \lambda(\ell) \chi(\ell)$ and $\lambda(\ell)$ are arbitrary complex numbers for $L < \ell \leq 2L$. Let χ_1 be the target harmonic. We choose $\lambda(\ell) = \bar{\chi}_1(\ell)$ to make $S(\lambda, \chi_1)$ large and realize the effect of amplification. Presumably the other sums $S(\lambda, \chi)$ with $\chi \neq \chi_1$ are, by orthogonality, quite small, but we do not need to show this as we can drop all of these terms by positivity. We get $L^2 |S(f, \chi_1)|^2 \ll LQ^2$, i.e. $S(f, \chi_1) \ll QL^{-1/2}$ provided L is a small power of Q, and this breaks the convexity bound.

Sometimes the amplifier cannot be as simple as $S(\bar{\chi}_1, \chi)$ because $\chi_1(\ell)$ vanishes quite often for small ℓ (see [**DFI3**] for other choices).

We finish this general presentation by showing one of many applications of the amplification technique to the L-function of the imaginary quadratic field $K = \mathbb{Q}(\sqrt{D})$. Let ξ be a character of the class group. Then we have (a work in progress by Duke–Friedlander–Iwaniec)

$$L_K(s, \xi) \ll |D|^{\frac{1}{4} - \frac{1}{2000}}$$

for any s with $\operatorname{Re} s = 1/2$, where the implied constant depends on s.

Remark. Some components of the amplification method, especially the positivity and completeness arguments, can be found, if one maintains an open mind, in certain techniques for estimating exponential sums in several variables ("glueing" and "smoothing" variables).

A lot of harmonic analysis is being applied in the recent works of N. Katz and P. Sarnak on the spacing of zeros of L-functions (cf. [**KS**]). A different analysis stemming from noncommutative geometry is being put to the test in the search for a proof of the Riemann hypothesis by A. Connes. Let us hope that some of these ideas will satisfy in a near future our present expectations!

References

[BFI] E. Bombieri, J. Friedlander and H. Iwaniec, *Primes in arithmetic progressions to large moduli*, Acta Math. 156 (1986), 203–251.

[Bir] B. J. Birch, *Forms in many variables*, Proc. Roy. Soc. Ser. A265 (1961/62), 245–263.

[Bom] E. Bombieri, *On the large sieve*, Mathematika 12 (1965), 201–225.

[Bru] R. W. Bruggeman, *Fourier coefficients of cusp forms*, Invent. Math. 45 (1978), 1–18.

[Bur] D. A. Burgess, *On character sums and L-series. II*, London Math. Soc. 13 (1963), 524–536.

[Cha] F. Chamizo, *Some applications of the large sieve in Riemann surfaces*, Acta Arith. 77 (1996), 315–337.

[DFI1] W. Duke, J. Friedlander and H. Iwaniec, *Equidistribution of roots of a quadratic congruence to prime moduli*, Annals of Math. 141 (1995), 423–441.

[DFI2] W. Duke, J. Friendlander and H. Iwaniec, *A quadratic divisor problem*, Invent. Math. 115 (1994), 209–217.

[DFI3] W. Duke, J. Friedlander and H. Iwaniec, *Class group L-functions*, Duke Math. J. 79 (1995), 1–56.

[DFI4] W. Duke, J. Friedlander and H. Iwaniec, *Representations by the determinant and mean values of L-functions*, in Sieve Methods, Exponential Sums, and their Applications in Number Theory, Cambr. Univ. Press, Cambridge 1997, 109–115.

[DI1] W. Duke and H. Iwaniec, *Convolution L-series*, Compositio Mathematica 91 (1994), 145–158.

[DI2] W. Duke and H. Iwaniec, *Estimates for coefficients of L-functions* I, in the Proceedings of the Conference on "Automorphic Forms and Analytic Number Theory" in Montreal 1989, McGill University, 43–47.

[DI3] W. Duke and H. Iwaniec, *Estimates for coefficients of L-functions* II, in the Proceedings of the Amalfi Conference on Analytic Number Theory in 1989, Università di Salerno 1992, 71–82.

[DRS] W. Duke, Z. Rudnick and P. Sarnak, *The density of integer points on an affine homogeneous variety*, Duke Math. J. 71 (1993), 143–180.

[Duk] W. Duke, *Hyperbolic distribution problems and half-integral weight Maass forms*, Invent. Math. 92 (1988), 73–90.

[DRS] W. Duke, Z. Rudnick and P. Sarnak, *The density of integer points on an affine homogeneous variety*, Duke Math. J. 71 (1993), 143–180.

[Duk] W. Duke, *Hyperbolic distribution problems and half-integral weight Maass forms*, Invent. Math. 92 (1988), 73–90.

[EMM] A. Eskin, G. A. Margulis and S. Mozes, *Upper bounds and asymptotics in a quantitative version of the Oppenheim conjecture*, to appear in Ann. Math., 1998.

[FI] J. Friedlander and H. Iwaniec, *The polynomial $X^2 + Y^4$ captures its primes*, to appear in Annals of Math.

[Fou] E. Fouvry, *Théorème de Brun–Titchmarsh; application au théorème de Fermat*, Invent. Math. 79 (1985), 383–407.

[Fri] J. Friedlander, *Bounds for L-functions*, in the Proceedings of the ICM in Zurich, 1995, Birkhauser Verlag, Basel 1995, 363–373.

[GK] S. W. Graham and G. Kolesnik, *Van der Corput's Method of Exponential Sums*, Cambridge University Press, Cambridge 1991.

[H-B] R. Heath-Brown, *A mean value estimate for real character sums*, Acta Arith. 72 (1995), 235–275.

[HR] G. R. Hardy and S. Ramanujan, *Asymptotic formulae in combinatory analysis*, Proc. London Math. Soc. 17 (1918), 75–115.

[Hej] D. Hejhal, *The Selberg Trace Formula for* PSL(2, R), Springer Lecture Notes in Math. 1001 (1983).

[Hux] M. N. Huxley, *Area, Lattice Points and Exponential Sums*, Oxford Science Publ., London 1996.

[I1] H. Iwaniec, *Small eigenvalues of Laplacian for* $\Gamma_0(N)$, Acta. Arith. 56 (1990), 65–82.

[I2] H. Iwaniec, *The lowest eigenvalue for congruence groups*, in Topics in Geometry, Birkhauser, Boston 1996, 203–212.

[I3] H. Iwaniec, *Fourier coefficients of modular forms of half-integral weight*, Invent. Math. 87 (1987), 385–401.

[I4] H. Iwaniec, *Introduction to the Spectral Theory of Automorphic Forms*, Bibl. Revista Mathem. Iberoamericana, Madrid, 1995.

[KS] N. M. Katz and P. Sarnak, *Random Matrices, Frobenius Eigenvalues, and Monodromy* (to appear).

[Kat] N. Katz, *Gauss Sums, Kloosterman Sums, and Monodromy Groups*, Annals of Math. Studies no. 116, Princeton University Press, Princeton 1988.

[KiSh] H. H. Kim and F. Shahidi, *Symmetric cube L-functions of* GL₂ *are entire*, Preprint 1997.

[Klo] H. D. Kloosterman, *On the representation of numbers in the form $ax^2 + by^2 + cz^2 + dt^2$*, Acta Math. 49 (1926), 407–464.

[Kuz] N. V. Kuznetsov, *Petersson's conjecture for cusp forms of weight zero and Linnik's conjecture. Sums of Kloosterman sums.* Math. USSR Sbornik 29 (1981), 299–342.

[LRS] W. Luo, Z. Rudnick and P. Sarnak, *On Selberg's eigenvalue conjecture*, Geom. and Funct. Analysis 5 (1995), 387–401.

[LS] W. Luo and P. Sarnak, *Quantum ergodicity of eigenfunctions on* PSL($\mathbb{Z}$)\$\mathbb{H}$, Public. Math. I.H.E.S. no. 81 (1995), 207–237.

[Lin1] Yu. V. Linnik, *Additive problems and eigenvalues of the modular operators*, in the Proc. ICM Stockholm 1962, 270–384.

[Lin2] Yu. V. Linnik, *The large sieve*, Dokl. Akad. Nauk USSR, 30 (1941), 292–294.

[Mic] P. Michel, *Autour de la conjecture de Sato–Tate pour les commes de Kloosterman I*, Invent. Math. 121 (1995), 61–78.

[Mon] H. Montgomery, *The analytic principle of the large sieve*, Bull. Amer. Math. Soc. 84 (1978), 547–567.

[PR]   R. Phillips and Z. Rudnick, *The circle problem in the hyperbolic plane*, J. Funct. Anal. 121 (1994), 78–116.

[Sar] P. Sarnak, *Arithmetic Quantum Chaos*, R. A. Blyth Lectures, University of Toronto, 1993.

[Sch] W. Schmidt, *The density of integer points on homogeneous varieties*, Acta Math. 154 (1985), 243–296.

[Tot] A. Toth, Thesis, Rutgers 1996

[Wey] H. Weyl, *Uber die Gleichverteilung von Zahlen mod Eins*, Math. Ann. 77 (1916), 313–352.

RUTGERS UNIVERSITY, DEPARTMENT OF MATHEMATICS, NEW BRUNSWICK, NJ 08903-2101
E-mail address: iwaniec@math.rutgers.edu

Symplectic Topology and Capacities

Dusa McDuff

I am going to talk about symplectic topology, a field that has seen a remarkable development in the past 15 years. Let's begin with what was classically known— which was rather little. We start with a symplectic form ω, that is, a closed 2-form which is nondegenerate. This last condition means that ω is defined on an even dimensional manifold M^{2n} and that its wedge with itself n times is a top dimensional form which never vanishes:

$$\omega \wedge \cdots \wedge \omega \neq 0 \qquad \text{everywhere.}$$

For example, in $\mathbb{R}^{2n}$ we can take

$$\omega_0 = \sum_{i=1}^{n} dx_i \wedge dy_i \tag{1}$$

where $x_1, y_1, \ldots, x_n, y_n$ are the coordinates of $\mathbb{R}^{2n}$.

The structure given by this basic form ω_0 first arose in Hamilton's formulation of classical mechanics in the mid 19th century, where he displayed the symplectic form as the mediator between the energy function and the time evolution equations of the system. More precisely, any function $H : \mathbb{R}^{2n} \to \mathbb{R}$ defines a vector field X_H via the equation:

$$\omega_0(X_H, \cdot) = dH(\cdot).$$

The flow ϕ_t generated by X_H is a family of transformations from $\mathbb{R}^{2n}$ to itself that has the following interpretation: for each point p in the phase space $\mathbb{R}^{2n}$, the image point $\phi_t(p)$ represents the state at time t of a system that was in state p at time 0. Moreover the number $H(p)$ is the energy of the system in state p. It is not hard to check that the transformations ϕ_t in this flow preserve the symplectic form, that is $\phi_t^*(\omega_0) = \omega_0$. Transformations (or diffeomorphisms) with this property are called *symplectomorphisms*. Obviously it is of great interest to understand their geometric properties, and, if possible, to find ways of distinguishing them from arbitrary smooth diffeomorphisms.

The first theorem in the subject is due to Darboux:

Any symplectic form ω on any manifold M^{2n} is locally diffeomorphic to ω_0.

In other words there are coordinates in a neighborhood of any point so that the form looks like (1). This is an example of a *local* to *global* phenomenon that occurs everywhere in symplectic topology. By this I mean the following. If first you

---

1991 *Mathematics Subject Classification.* 53C15.

look linearly, at a tangent space for example, then up to a linear coordinate change there is only one linear nondegenerate 2-form, namely (1). Now, Darboux's theorem tells you that the uniqueness that you have at a point extends to uniqueness in a neighborhood. One can also interpret this in terms of a passage from linear to nonlinear: a linear phenomenon persists in the nonlinear situation.[1]

The second basic theorem is Moser's stability theorem concerning families of forms. Suppose we have a family of symplectic forms ω_t, $0 \le t \le 1$, on some closed manifold (compact without boundary.) Let's also suppose that the cohomology class $[\omega_t] \in H^2$ of these forms is fixed, so that the variation is just by exact forms: $\omega_t = \omega_0 + d\sigma_t$. Then the theorem is that the forms are all basically the same. That is, there exists a family of diffeomorphisms ψ_t, $0 \le t \le 1$, of the manifold starting at the identity (that is $\psi_0 = \mathrm{id}$) such that $\psi_t^* \omega_t = \omega_0$ for all t.

What this says is there is no interesting deformation theory of symplectic forms: if you try to change the forms within a fixed cohomology class you can't do it. The proof involves finding a (time-dependent) vector field X_t which integrates to give the isotopy ψ_t. For this to work the manifold must be closed, since otherwise the flow ψ_t might run off it at infinity. However, there is an extension of this theorem to open manifolds if you put in some controls at infinity.

So that is what was known: not much at this point. The story that I really want to tell you begins with Gromov's work [9] in the mid 80s where he introduced elliptic methods into symplectic topology and proved a whole array of wonderful results.[2] First I shall state some of them and then give you an idea of how to prove them.

Gromov's results

(I) The first result is known as the **nonsqueezing theorem**. The basic question here is to understand the possible shapes of the image of a standard ball under a symplectic transformation. Gromov picked out as decisive the question of when a ball can be symplectically mapped inside a cylinder. Here the ball is the standard (compact) ball of radius r sitting in standard Euclidean space, $B^{2n}(r) \subset (\mathbb{R}^{2n}, \omega_0)$, and the cylinder $B^2(\lambda) \times \mathbb{R}^{2n-2}$ is the product of a 2-disc of radius λ with Euclidean $(2n-2)$-space. It is important here that the latter is a symplectic product; that is the symplectic form on the first factor is the area form $dx_1 \wedge dy_1$, and that on the second is the basic form in the remaining variables.

Gromov's theorem tells us that there is such a symplectic embedding if and only if $r \le \lambda$:

$$B^{2n}(r) \hookrightarrow B^2(\lambda) \times \mathbb{R}^{2n-2} \quad \Longleftrightarrow \quad r \le \lambda. \tag{2}$$

Obviously if $r \le \lambda$, you can just include the ball inside the cylinder. The force of this theorem is that, if $r > \lambda$, it is impossible to take the round ball of radius r and squeeze it symplectically to fit it into the cylinder, no matter how long you make it. This is a very clean statement of what a symplectic map cannot do, and shows there is some kind of fatness in the ball that you can't squeeze.[3]

---

[1] For proofs of this and other results, and further references, the interested reader can consult McDuff and Salamon [20].

[2] Another important theme in the development of symplectic topology came with Conley and Zehnder's use of variational methods in their proof in 1983 of Arnold's conjecture for the torus: see [3] and Hofer–Zehnder's book [10].

[3] Lalonde and McDuff [12] have recently shown that a similar result is true for embeddings $B^{2n}(r) \to B^2(\lambda) \times M^{2n-2}$ for any symplectic manifold M, closed or not.

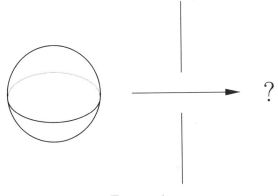

FIGURE 1

Observe also that this property distinguishes symplectic from volume-preserving transformations. Every symplectic map preserves volume, since volume is given by the form $\omega_0^n = \omega_0 \wedge \cdots \wedge \omega_0$. Moreover, it is not hard to see that one can map *any* ball into *any* cylinder if all one requires is that the volume form is preserved, since maps of the form

$$(x_1, y_1, x_2, y_2, \dots) \mapsto (\mu x_1, \mu y_1, \mu^{-1} x_2, \mu^{-1} y_2, \dots)$$

preserve volume. Thus the nonsqueezing theorem shows that symplectic transformations are much more limited than volume-preserving ones. In particular, it is not possible to approximate an arbitrary volume-preserving transformation by a symplectic one in the uniform (or C^0) topology. Incidentally, one can also think of this result as another instance of the linear to nonlinear phenomenon: it is very easy to check that there is no *linear* (or affine) symplectic transformation of Euclidean space that squeezes a ball into a thinner cylinder, and now we see that this cannot be done by an arbitrary nonlinear symplectic transformation either.

(II) Another very nice result is the solution of the "camel" problem. This problem is the following: imagine you are in Euclidean space $\mathbb{R}^{2n}$ minus a wall W (Figure 1).

This wall consists of the hyperplane $x_1 = 0$ with a hole of radius 1 in it and is defined by the equations

$$x_1 = 0 , \quad \sum_{i>1} x_i{}^2 + \sum_{i \geq 1} y_i{}^2 \geq 1.$$

Now imagine putting a big ball $B^{2n}(r)$, $r > 1$, on the left side of Figure 1, and ask: is there a way of deforming this ball symplectically so as to take it through the hole and over to the other side? One might first of all see whether it is possible to go through the hole preserving volume. But now you don't see the roundness in the ball: you can make the ball long and thin preserving volume, and then slide it through. The nonsqueezing theorem says that this manoever is impossible symplectically. However it still might be possible to get the ball through the hole in some other more complicated way—all we need now is to squeeze the ball in one hyperplane. Nevertheless, as you might guess, the answer to this problem is no. There is some sort of obstruction that prevents the ball from going through.

(III) Another result in which people are very interested these days is the uniqueness of the symplectic structure on $\mathbb{R}^4$ and $\mathbb{CP}^2$. The preceding results work in all dimensions but this one is definitely 4-dimensional. What Gromov proved is the following. Suppose we have a 4-dimensional symplectic manifold (X, ω) which, outside a compact set, is symplectomorphic to $\mathbb{R}^4$ minus a ball. Then, provided that $\pi_2 X = 0$, this symplectomorphism extends to a symplectomorphism from the whole of X onto $\mathbb{R}^4$. Note that the conclusion includes in it a statement of what the diffeomorphism type of X is, while all that is assumed is some homotopic theoretic knowledge and information about the symplectic structure on X at infinity.

The above result is actually equivalent to a uniqueness result for $\mathbb{CP}^2$, that can be stated as follows. Suppose you have a compact symplectic manifold (Z, ω) whose second homotopy group $\pi_2(Z)$ is generated by a symplectically embedded 2-sphere S with self intersection $+1$. (This means that $S \cdot S = 1$ or, equivalently, that the normal bundle to S has Chern number 1.) Then this manifold (Z, ω) is symplectomorphic to $\mathbb{CP}^2$ with its standard symplectic structure, appropriately scaled. To see the equivalence, observe that if you remove a neighborhood of the sphere S from Z you get a manifold whose boundary is symplectomorphic to the standard 3-sphere in $\mathbb{R}^4$ and which can therefore be extended to a symplectic manifold X that looks like $\mathbb{R}^4$ at infinity. Conversely, we can compactify X by removing a standard collar at infinity and then attaching a 2-sphere S to produce such a Z.

Note that Gromov can only establish uniqueness under the assumption that the manifold Z contains a symplectic 2-sphere S. One of the triumphs of the recent work of Taubes [25] on Seiberg–Witten theory is to remove this assumption. Thus one now knows that a symplectic 4-manifold that is diffeomorphic to $\mathbb{CP}^2$ is actually symplectomorphic to $\mathbb{CP}^2$.

(IV) As a counterpoint to the above uniqueness results, Gromov showed that when $n > 1$ there exists an exotic symplectic structure on $\mathbb{R}^{2n}$. In other words, there is a structure ω that does not live in standard Euclidean space$(\mathbb{R}^{2n}, \omega_0)$ in the sense that there is no embedding

$$\psi : \mathbb{R}^{2n} \to \mathbb{R}^{2n}$$

such that $\psi^*(\omega_0) = \omega$. Here the difficulty is not so much in constructing something that you think should be exotic, but in proving that it is exotic.

Gromov's criterion hinges on properties of Lagrangian submanifolds of Euclidean space. An n-dimensional submanifold L of a symplectic manifold (M^{2n}, ω) is called Lagrangian if the restriction of ω to L is identically zero. This implies that if ω is exact (i.e. $\omega = d\lambda$), the 1-form λ restricts to a closed form λ_L on L since $d\lambda_L = \omega|_L = 0$. Hence λ defines a de Rham cohomology class $[\lambda_L] \in H^1(L)$. If in turn this class $[\lambda_L]$ is zero (or, equivalently, if $\lambda|_L$ itself is exact, $\lambda_L = dF$), the Lagrangian submanifold L is said to be *exact*. Gromov's main result is that *no* closed Lagrangian submanifold in the standard $(\mathbb{R}^{2n}, \omega_0)$ is exact. He then constructed a symplectic structure ω on $\mathbb{R}^{2n}$ that contained an exact closed Lagrangian submanifold L, and concluded that ω had to be exotic since otherwise L would give rise to a forbidden exact Lagrangian submanifold in standard Euclidean space.

Nobody has yet managed to do much more with this problem, for example, showing that there is more than one exotic structure on Euclidean space, or finding an exotic structure that is standard at infinity (by (III) above this would have to live in dimension at least 6.)

J-holomorphic curves

I'll now try to give you an idea of how Gromov proved these results. Observe first that if you have a symplectic manifold (M, ω) then you can always find an almost complex structure J on the manifold. That's an automorphism of the tangent bundle TM which, like multiplication by i, satisfies the equation $J^2 = -\mathrm{Id}$. Moreover you can require this almost complex structure to be related to ω by the taming (or positivity) condition:

$$\omega(v, Jv) > 0, \quad \text{whenever} \quad v \in T_p, \ v \neq 0, \tag{3}$$

so that there is an associated Riemannian metric on M, given by the symmetrization of $\omega(v, Jv')$. It is easy to check that the set of J satisfying these conditions (these are called ω-tame J) is nonempty and contractible. Therefore one can try to find invariants of (M, ω) by looking for invariants of the almost complex manifold (M, J) that do not depend on the choice of ω-tame J.

Gromov constructed his invariants by looking at maps of Riemann surfaces[4] into the almost complex manifold

$$u : (\Sigma, j) \to (M, J)$$

which satisfy the generalized Cauchy–Riemann equation:

$$du \circ j = J \circ du. \tag{4}$$

Such maps u are called J-holomorphic curves. Because equation (4) is elliptic, its solution spaces have very nice properties. For example, for generic ω-tame J, the space of solutions in a fixed homology class A is a finite dimensional manifold $\mathcal{M}(A, J)$. It's not true that these solution (or moduli) spaces are compact because curves can degenerate. But the positivity condition (3) allows one to control and understand the degenerations that occur and hence describe the compactified moduli spaces of curves. Moreover, because the space of ω-tame almost complex structures J is path-connected, any two generic J can be connected by a path J_t such that the set of all J_t-holomorphic curves for $t \in [0, 1]$ forms a cobordism between the moduli spaces at $t = 0, 1$. In many situations, this cobordism is compact, which means that any property of the moduli space $\mathcal{M}(A, J)$ which is a cobordism invariant depends only on the underlying symplectic form ω and not on J. For example, the Gromov invariants are defined by counting the number of J-holomorphic curves (with given genus and in a given homology class) that go through a fixed number of points or cycles.

I'll now show how these ideas can be used to prove the nonsqueezing theorem. Consider the ball $B^{2n}(r)$ with its standard complex structure J_0. First observe that if you have a J_0-holomorphic curve S that goes through the center of the ball and is properly embedded (so it goes all the way to the boundary) then the symplectic area of S is at least πr^2. In fact, J_0-holomorphic curves in $\mathbb{R}^{2n}$ are just complex curves in the usual sense, and so are minimal surfaces with respect to the usual metric g_0. Moreover, it is easy to see that the symplectic area $\int_S \omega_0$ of a complex curve S is just its area with respect to the usual metric. Hence the above result holds because of the well known fact that the surface of minimal area through the center of $B^{2n}(r)$ is the flat disc, which has an area πr^2.

So now take this ball and suppose it is embedded in a cylinder. On the image you have the pushforward of J_0, which you can always extend to an ω_0-tame J that

---

[4]A Riemann surface is a compact 2-manifold Σ with complex structure j.

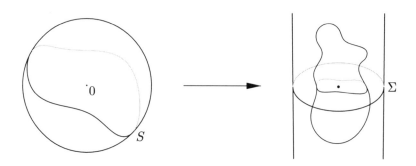

FIGURE 2

is standard near the boundary of the cylinder. Then J extends to the compact manifold $S^2 \times \mathbb{R}^{2n-2}$ obtained by closing up the cylinder. Now, if J is the product almost complex structure on $S^2 \times \mathbb{R}^{2n-2}$ there is a flat J-holomorphic 2-sphere through every point, that is unique modulo reparametrization. Moreover one can show that this product J is generic in the requisite sense (that is, it is a regular point for an appropriate operator). Hence the Gromov invariant that counts the number of spheres in the class $[S^2 \times \mathrm{pt}]$ through a fixed point takes the value $+1$. The theory outlined above implies that this number is independent of the choice of J. Therefore, for any ω-tame J on the product and any point p, there is precisely one J-holomorphic 2-sphere through p in the homology class of $S^2 \times \mathrm{pt}$, when these are counted with appropriate signs and modulo parametrization. In particular, for our special J which equals the pushforward of the standard structure on the image of the ball, there is at least one J-holomorphic sphere Σ going through the image of the center of the ball. Note that the symplectic area of Σ is determined by its homology class and so is $\pi(\lambda + \epsilon)^2$ for some arbitrarily small $\epsilon > 0$. (This ϵ appears because we slightly increase the size of the cylinder when we close it up.) Now look at the inverse image S of Σ in the ball. (See Figure 2).

By our previous result, the area of S is at least πr^2. Moreover, since symplectomorphisms preserve area, this has to be strictly less than the area of Σ, which is $\pi(\lambda + \epsilon)^2$. Since this is true for all $\epsilon > 0$ we find that $r \leq \lambda$.

A similar but more elaborate version of this idea proves the camel theorem: one just shows that there are a lot of J-holomorphic disks with boundaries on the hole and area at most π that give obstructions to putting the ball through. Finally, for the uniqueness of structure theorem one shows that there are so many spheres that one can somehow use them as coordinates to construct a diffeomorphism from one manifold to another. The argument about Lagrangian submanifolds L is somewhat more complicated but it also uses the behavior of J-holomorphic curves (in this case, discs with boundary on L) as its technical base.

Since their introduction in the mid-80s these elliptic methods have been very influential, especially when combined with Floer's approach to Morse theory: see for example Floer [7]. A detailed description of the construction of genus 0 Gromov invariants, together with a discussion of their relation to quantum cohomology and problems in enumerative geometry, may be found in Ruan–Tian [22] and McDuff–Salamon [19]. The higher genus case is treated in Ruan–Tian [23].

Capacities

One set of applications of Gromov's ideas revolves around the notion of symplectic capacity that was formalized by Ekeland and Hofer [6]. A symplectic capacity c is a function of subsets in Euclidean space (or more generally, in arbitrary symplectic manifolds of a fixed dimension) taking values in $[0, \infty]$, and is a measure of how big this subset is. It satisfies these conditions:

(i) If $(U, \omega) \subset (V, \omega)$ then $c(U, \omega) \leq c(V, \omega)$, that is, c is monotonic.

(ii) If $\phi : (U, \omega) \to (V, \omega')$ is a diffeomorphism which is conformally symplectic, that is $\phi^*(\omega') = \lambda \omega$ for some constant λ, then $\lambda c(U, \omega) = c(V, \omega')$.

(iii) The third condition is a normalization condition which specifies that the capacity of a ball with its standard structure is positive: $c(B^{2n}(r)) > 0$.

(iv) Finally you need something which tells you that you are not measuring a volume; you want something that is definitely two dimensional. This is formulated as: $c(B^2(r)) \times \mathbb{R}^{2n-2} < \infty$.

An example of capacity is this:

$$c_G(M) = \sup \left\{ \pi r^2 : B^{2n}(r) \hookrightarrow M \text{ symplectically} \right\}$$

Here you are just measuring the size of the largest ball which embeds symplectically in M. This obviously satisfies (i) and (ii). Condition (iii) follows from Darboux's theorem: there are small symplectic balls in M. Finally, the non-squeezing theorem gives (iv):

$$c_G(B^2(\lambda) \times \mathbb{R}^{2n-2}) = \pi \lambda^2$$

I shall call this capacity defined by embedding balls the *Gromov capacity c_G*. There are other symplectic capacity functions developed by Ekeland–Hofer and Hofer–Zehnder (for precise references see [20, 10]) that arise from variational problems, but I won't go into those.

One important use of symplectic capacities, noted by Eliashberg [5] and by Ekeland and Hofer, is to give a topological criterion for deciding whether a diffeomorphism is symplectic or not.[5] Let's suppose that you have a capacity (such as c_G) with the specific normalization condition that the capacity of the ball of radius R is the same as that of the cylinder of radius R:

$$c(B^{2n}(R)) = c(B^2(R) \times \mathbb{R}^{2n-2}).$$

Then the statement is:

A diffeomorphism ϕ of $\mathbb{R}^{2n}$ is symplectic if and only if ϕ preserves capacity: that is, $c(\phi(W)) = c(W)$ for all open subsets W of $\mathbb{R}^{2n}$.

The proof of this statement is based on the fact that capacity is invariant under rescaling. In order to establish that a map ϕ is symplectic one just has to show that its derivative at each point is a linear symplectic map. But when you take a derivative at a point you look at a little piece of space and magnify it until in the limit you get the derivative. Thus taking the derivative is just the limit of a rescaling process, and so it is not hard to see that if a map ϕ preserves capacity

---

[5]This criterion is called topological because it makes no use of the derivative, and hence implies that the group of symplectomorphisms is closed in the group of all diffeomorphisms with respect to the uniform topology.

its derivative also does. Now it is just a matter of linear algebra to show that the resulting linear map is symplectic: in fact, any linear map L that preserves capacity is symplectic or antisymplectic,[6] and it is not hard to rule out the antisymplectic case.

The symplectic topology of Euclidean space

Because Darboux's theorem shows that any symplectic form is locally like the standard structure in Euclidean space, there is a temptation to to think that there is nothing to be said about local symplectic geometry, all forms are locally the same and that is that. But it turns out that the standard symplectic structure has a whole lot of interesting properties.

Here I'll mention some things that can be proved using the idea of capacity. First, consider an open polydisc $P(r_1, r_2, \ldots, r_n)$—the product of n open 2-dimensional discs of radii $r_1, \ldots, r_n$, with normalization condition $r_1 \leq r_2 \leq \cdots \leq r_n$. Gromov asked: are these symplectic manifolds (with the standard structure inherited from Euclidean space) all different? The point of his question is this. A closed (smooth) domain in Euclidean space has a characteristic Hamiltonian flow on its boundary and you can easily distinguish among closed polydiscs by looking at this boundary flow.[7] If you look at open polydiscs you don't have that flow, but you might suspect that the flow on the boundary would leave enough trace inside to force the open domains to be different.

In fact, when $n = 2$ it is easy to show that this suspicion is true by using the Gromov capacity. Since

$$c_G(P(r_1, r_2)) = \pi r_1^2, \qquad \text{vol}\,(P(r_1, r_2)) = \pi^2 r_1^2 r_2^2,$$

one can distinguish two 4-dimensional polydiscs using these two invariants. In other words, $P(r_1, r_2)$ is symplectomorphic to $P(r_1', r_2')$ if and only if $r_1 = r_1', r_2 = r_2'$. But when $n \geq 3$ these two invariants are no longer enough. However Floer and Hofer developed a theory of symplectic homology (a kind of Floer homology built out of action functionals coming from Hamiltonians) that extends the notion of capacity and, using this, Floer–Hofer–Wysocki showed in [8] that the polydiscs are all distinct. A similar result holds for ellipsoids. These methods can also be used to throw light on the question of when one ellipsoid or polydisc may be symplectically embedded in another. For very nice recent work on this topic see Schlenk [24].

Another kind of question is this: Consider the set $\text{Emb}(B^{2n}(r), U)$ of all symplectic embeddings of a ball into the set U—here U could be a domain in $\mathbb{R}^{2n}$ or any symplectic manifold. Capacity tells you the maximum size of a ball that you can embed, but now you can ask whether this space of embeddings is connected. This can be thought of as an extension of the camel problem, because the camel problem tells you that if the ball embedded on one side of a wall with a hole has radius larger than that of the hole, this space is disconnected. One could ask:

---

[6]That is, $L^*(\omega_0) = \pm\omega_0$.

[7]This flow is simply the Hamiltonian flow—generated by the vector field X_H defined earlier—of a smooth function H that is constant on the boundary. The fact that a closed polydisc has nonsmooth boundary causes no real problem here. Any symplectomorphism ϕ that takes one closed polydisc P to another P' maps a sequence of smooth approximations to the boundary ∂P into a similar approximating sequence for $\partial P'$. But a symplectomorphism defined only on an open polydisc need not extend to the boundary and so need not take an approximating sequence to $\partial P'$ into an approximating sequence for $\partial P'$.

FIGURE 3

> *For which domains in Euclidean space is this embedding space disconnected?*

Another interesting question to which I don't have an answer is this:

> *Is there a closed manifold M such that this embedding space is disconnected?*

It is likely that the answer depends on r: the embedding space might always be connected for small enough r but then become disconnected for large r, much as happens in the camel problem. So far, the only results on this question are in dimension 4. For example I showed using J-holomorphic curves that this space is always connected when the target U is a 4-ball or more generally a star-shaped region. There is also one relevant result of Floer–Hofer–Wysocki that considers what happens when the domain and target spaces both have corners: they prove that $\mathrm{Emb}(P(r_1, r_2), P(1, 1))$ is disconnected if $r_1, r_2 < 1$ but $r_1^2 + r_2^2 > 1$. The two embeddings which are not isotopic are the obvious inclusions (see Figure 3).

You see that to get from one image to the other you have to rotate a quarter turn. However, using symplectic homology one can show that if $r_1^2 + r_2^2 > 1$ the corners of the domain are too large for it to be able to be turned in this way. (The corners are detected using products in the homology theory.)

Symplectic deformations and isotopies

I have now told you something of what is known about the symplectic topology of Euclidean space. Now, I would like to talk about symplectic structures on general manifolds. Remember that Moser's stability theorem says that if you have a family of symplectic forms ω_t, $0 \leq t \leq 1$, on a closed manifold M with constant cohomology class then this family is just gotten by a isotopy ψ_t of the underlying manifold. Thus $\psi_t^*(\omega_t) = \omega_0$ for all t. Such a family ω_t is called an *isotopy*. There is another related notion where you have a family of symplectic forms as above, but you don't make any assumption on the cohomology class, instead allowing it to vary. That is called a *deformation*. And you might ask:

> *Suppose you have a deformation $\{\omega_t\}$ such that $[\omega_0] = [\omega_1]$, is the family homotopic through deformations with fixed endpoints to an isotopy?*

That is, you ask if you can get anything new in a given cohomology class if you allow yourself to make deformations.

Examples in 6 dimensions

There are certain examples which tell you that in dimension six and above deformations are different from isotopies. I can just briefly describe them. Take $M = S^2 \times S^2 \times T^2$ and let the symplectic form

$$\omega = \sigma_0 \oplus \sigma_1 \oplus \sigma_2, \tag{5}$$

be the sum of the area forms on each factor. What is very important is that you insist that the area of the first factor is the same as the area of the second factor. Suppose you take a diffeomorphism ϕ which twists the first factor as follows. Let (z, w, s, t) be the coordinates for M, where $z, w \in S^2, s, t \in T^2 = \mathbb{R}^2/\mathbb{Z}^2$. Then define

$$\phi(z, w, s, t) = (z, R_{z,t}w, s, t)$$

where $R_{z,t}$ is rotation of the sphere by $2\pi t$ about the axis through z. (Here we think of the 2-sphere S^2 embedded in 3-space $\mathbb{R}^3$ in the usual way so that the axis through z is just the radius $0z$.) Then the claim is that the pull back $\phi^*\omega_0$ of ω_0 by this twisting ϕ is deformation equivalent to ω_0, but not isotopic to it. Interestingly enough, if you put a constant $\lambda > 1$ in (5)

$$\omega^\lambda = \lambda\sigma_0 \oplus \sigma_1 \oplus \sigma_2, \tag{6}$$

so that the first sphere is bigger than the second one, then the forms $\phi^*(\omega^\lambda)$ and ω^λ are actually isotopic. (This immediately implies that $\phi^*(\omega)$ and ω are deformation equivalent.) But when $\lambda = 1$ you can't undo the effect of the twisting. In fact, you can see this twisting symplectically by looking at the family of spheres $z \mapsto (z, w_0, s_0, t)$ for $t \in S^1$. These spheres are J-holomorphic when J is the product almost complex structure J_{split}, and persist as J varies among all ω-tame almost complex structures. In other words for each such J one has a map $S^2 \times S^1 \to M$ with image equal to a family of J-holomorphic spheres. Moreover, for each J, the composite of this map with the projection pr_2 onto the second sphere is bordant to a constant map.[8] However, if one looks at J that are tamed by the twisted form $\phi^*(\omega)$ the story is different. For example if one takes $J = \phi^*(J_{\text{split}})$ the family is

$$z \mapsto \phi^{-1}(z, w_0, s_0, t) = (z, R_{z,-t}w_0, s_0, t),$$

and so, composition with projection onto the second sphere gives the map

$$e_J : S^2 \times S^1 \to S^2, \qquad (z, t) \mapsto R_{z,-t}(w_0).$$

Using a version of the Hopf invariant, one easily sees that this map is not bordant to a constant map. A similar map e_J can be defined for any J that is tamed by a form isotopic to $\phi^*(\omega)$, and it follows from the general theory that all such maps are bordant. Hence ω cannot be isotopic to $\phi^*(\omega)$. The argument breaks down when $\lambda > 1$ since in this case the moduli space of curves we are considering fails to be compact. This implies that our method only gives a noncompact bordism

---

[8]Two maps $f_i : X_i \to Y$ are said to be bordant if there is a manifold W with boundary equal to $X_0 \cup X_1^{\text{opp}}$ and a map $g : W \to Y$ such that g restricts to f_i on X_i for $i = 0, 1$. Here I am assuming that all the manifolds X_i, W are oriented and compact and am writing X^{opp} for X with the opposite orientation. For example, homotopic maps are bordant, though not conversely.

$g : W \to S^2$ between the maps e_J as J varies. Since all maps are noncompactly bordant, the invariant disappears.

I would also like to mention that there are some examples of Ruan [21] which give non deformation equivalent symplectic structures on 6 dimensional manifolds of the form $M^6 = X^4 \times S^2$. The idea is that you take two non diffeomorphic but homotopy equivalent symplectic four manifolds X, Y. When you stabilize them (cross with S^2) they become diffeomorphic. However, the pattern of J-holomorphic curves they contain persists under stabilization and therefore they are often symplectically distinct. For example, one can take X equal to the complex projective plane blown up at 8 points and Y equal to the Barlow surface. Then $X \times S^2$ contains 8 families of J-holomorphic spheres corresponding to the blown up points, while $Y \times S^2$ contains no J-holomorphic spheres except for those in class [pt $\times S^2$].

The 4-dimensional case

Finally, something about 4-dimensional manifolds. As mentioned, Taubes–Seiberg–Witten theory allows a significant strengthening of Gromov's uniqueness results. You see, Gromov had to assume in his statement that some structure was known—either there was a symplectically embedded 2-sphere or in the case of Euclidean space the structure was known at infinity. But now this is unnecessary, at least in the compact case, since Taubes has shown that the Seiberg–Witten invariants of a closed symplectic 4-manifold do not vanish, and that nonvanishing invariants produce symplectically embedded curves. Together with some work of Li–Liu [15], Liu [16] and Lalonde–McDuff [13], this completely settles questions on the symplectic structure of projective space and ruled manifolds: for a survey and detailed references see Lalonde–McDuff [14].

Ruan's 6-dimensional examples mentioned above have the property that the two symplectic forms are not deformation equivalent even after pull back by an arbitrary diffeomorphism. It is not known in 4 dimensions whether such examples can exist. In other words one can ask:

> Is there a 4-manifold X that supports two symplectic forms ω_0 and ω_1 with the property that $\psi^*\omega_1$ is not deformation equivalent to ω_0 for any diffeomorphism ψ?

Another question is:

> Are two deformation equivalent and cohomologous forms ω_0, ω_1 on a 4-manifold necessarily isotopic?

The point here is that isotopy is the basic relation if you are interested in the symplectic geometry of a manifold, but the Gromov invariants defined by counting J-holomorphic curves are invariant under deformation. In fact, in 4-dimensions Taubes has shown that the Gromov invariants coincide with the Seiberg–Witten invariants and so depend only on the smooth structure. (For more information on different ways of counting curves see Ionel–Parker [11].) The 6-dimensional examples of deformation equivalent but nonisotopic forms used an invariant of the moduli space that was more complicated than simply counting the numbers of curves through a set of points (or, more generally, through some k-cycle.) Therefore, one might hope that there would be symplectic 4-manifolds X with positive dimensional moduli spaces of curves that one could use to create new invariants fine enough to distinguish some nonisotopic forms. In order to get an invariant, these moduli spaces would have to be nonempty for generic J, and the easiest

way to ensure this is to ask that the corresponding Gromov invariant is nonzero.[9]
In the simplest case, these moduli spaces would appear as solution spaces to the
Seiberg–Witten equations for some Spin^c structure, and so X would have to have
$b_2^+ = 1$ and be of so-called nonsimple SW-type. Such manifolds do exist: for example blow-ups of ruled surfaces or of the complex projective plane. However, there
is an inflation technique due to Lalonde–McDuff (see [13]) that allows one to use
an embedded J-holomorphic curve to change the cohomology class of a family of
symplectic forms, and if there are enough such curves one can alter this cohomology
class almost at will. Hence one can show:

> If X is a symplectic manifold of nonsimple SW-type, any deformation
> ω_t between two cohomologous symplectic forms ω_0 and ω_1 is homo-
> topic through deformations with fixed endpoints to an isotopy.

This rules out any easy way of finding nonisotopic but cohomologous symplectic
forms on a 4-manifold. It also allows one to answer a question about the uniqueness
of symplectic blowing up that I have been thinking about for quite some time. When
you blow up a symplectic manifold you embed a lot of disjoint balls, cut out their
interiors and then glue up their boundaries: for details see [20]. The cohomology
class of the blow up is determined by the size of the balls. Moreover, in dimension
4, isotopy classes of forms on the blow up are in bijective correspondence with path
components of the relevant space of embeddings of balls. Since these embedding
spaces are clearly path connected if the balls can be reduced in size, it is easy to
see that any two blow up forms are deformation equivalent. Therefore,

> If X satisfies the hypotheses above, in particular if X is a rational or
> ruled surface, cohomologous blow ups are unique up to isotopy.

Details of this argument may be found in McDuff [18]. One can also use similar
ideas to construct certain blow ups thereby solving the symplectic packing problem
for these manifolds: see Biran [2].

   In conclusion I would like to mention Donaldson's work [4] (see also Auroux
[1]) that constructs geometrically interesting "almost J-holomorphic" sections of
"sufficiently positive" line bundles on a symplectic manifold (of any dimension). In
particular, he has shown that if the cohomology class of the form ω is integral, it is
possible to represent a sufficiently large multiple of the Poincaré dual of this class
by a codimension 2 symplectic submanifold that is well-defined up to isotopy. This
promises both to give interesting obstructions to the existence of symplectic forms
in dimensions > 4 and to yield many more invariants of symplectic manifolds. In
particular, one may be able to use it to prove Donaldson's stabilization conjecture
which says that if X, Y are symplectic 4-manifolds with the same homotopy type
then the obvious symplectic forms on $X \times S^2$ and $Y \times S^2$ are deformation equivalent
if and only if X and Y are diffeomorphic. If something like this were true, then
proceeding inductively one might find that deformation classes of symplectic manifolds of any dimension larger than 4 are as complicated as diffeomorphism classes
of symplectic 4-manifolds.

References

[1] D. Auroux, Asymptotically holomorphic families of symplectic submanifolds, preprint
     (1997).

---

[9]It is no good taking X to be something like the 4-torus T^4 that has families of J-holomorphic
tori of real dimension 2 for some special J—a product, for example—but no tori for other J.

[2] P. Biran, Symplectic packing in dimension 4, preprint (1996).

[3] C. Conley and E. Zehnder, The Birkhoff–Lewis fixed point theorem and a conjecture of V.I. Arnold, *Inventiones Math.* **73** (1983), 33–49.

[4] S. K. Donaldson, Symplectic submanifolds and almost-complex geometry, *Journ. of Differential Geometry*, **44** (1996), 666–705.

[5] Y. Eliashberg, A theorem on the structure of wave fronts and its applications in symplectic topology, *Functional Analysis and Applications*, **21** (1987), 65–72.

[6] I. Ekeland and H. Hofer, Symplectic topology and Hamiltonian dynamics. *Mathematische Zeitschrift*, **200** (1989), 355–78.

[7] A. Floer, Symplectic fixed points and holomorphic spheres. *Communications in Mathematical Physics*, **120** (1989), 575–611.

[8] A. Floer, H. Hofer and K. Wysocki, Applications of symplectic homology I. *Mathematische Zeitschrift* **217** (1994), 577–606.

[9] M. Gromov, Pseudo holomorphic curves in symplectic manifolds, *Inventiones Mathematicae*, **82** (1985), 307–47.

[10] H. Hofer and E. Zehnder, *Symplectic capacities and Hamiltonian Dynamics*, (1994) Birkhauser, Basel.

[11] E. Ionel and T. Parker, The Gromov invariants of Ruan–Tian and Taubes, preprint (1997).

[12] F. Lalonde and D. McDuff, The geometry of symplectic energy, *Annals of Mathematics*, **141** (1995), 349–371.

[13] F. Lalonde and D McDuff, The classification of ruled symplectic 4-manifolds, *Math. Research Letters* **3** (1996), 769–778.

[14] F. Lalonde and D McDuff, *J*-curves and the classification of rational and ruled symplectic 4-manifolds, in *Symplectic and Contact Geometry*, ed. C. Thomas, Camb Univ Press (1996).

[15] T.J. Li and A. Liu, General wall crossing formula, *Math. Research Letters* **2** (1995), 797–810.

[16] A. Liu, Some new applications of general wall crossing formula, *Math. Research Letters*, **3** (1996), 569–586.

[17] D. McDuff, Examples of symplectic structures, *Inventiones Mathematicae*, **89**, (1987) 13–36.

[18] D. McDuff, From symplectic deformation to isotopy, to appear in the Proceedings of the conference at Irvine 1996, ed Stern, International Press.

[19] D. McDuff and D. Salamon, *J-holomorphic curves and quantum cohomology*, University Lecture Series, American Mathematical Society, Providence, RI. (1994)

[20] D. McDuff and D. Salamon, *Introduction to Symplectic Topology*, OUP, Oxford, UK (1995)

[21] Y. Ruan, Symplectic topology on algebraic 3-folds, *Journal of Differential Geometry* **39** (1994), 215–227.

[22] Y. Ruan and G. Tian, A mathematical theory of quantum cohomology, *Journal of Differential Geometry* **42**, (1995), 259–367.

[23] Y. Ruan and G. Tian, Higher genus symplectic invariants and sigma model coupled with gravity, *Inventiones Mathematicae* **130** (1997), 455–516.

[24] F. Schlenk, On symplectic folding, in preparation.

[25] C. H. Taubes, The Seiberg–Witten and the Gromov invariants, *Math. Research Letters* **2** (1995), 221-238.

SUNY at Stony Brook, Institute for Mathematical Sciences, Mathematics Building, Stony Brook, NY 11794-3660

E-mail address: dusa@math.sunysb.edu

Evolution Problems in Geometry and Mathematical Physics

Michael Struwe

ABSTRACT. A survey of recent results on geometric evolution problems and
flow equations in mathematical physics reveals some unifying features and
raises challenging questions.

1. Introduction

Curvature-driven motion is ubiquitous in geometry and mathematical physics.
The corresponding nonlinear partial differential equations pose challenging analytical problems. Regarding the issue of global well-posedness, that is, global existence
and uniqueness of (suitable) solutions, recent progress on these equations seems to
reveal a uniform pattern.

For better motivation we start with a problem that can be visualized easily.
Consider an embedded surface in space contracting under the influence of surface
tension. Such a surface then moves in direction of its mean curvature vector.
Surface energy, for instance, seems to be governing the motion of grain boundaries
in annealing metal; we thus expect these interfaces to be moving by mean curvature.

Analytically, mean curvature flow is described by a family of smooth embeddings $F(t, \cdot), t \geq 0$, of an m-dimensional model hypersurface M into an n-
dimensional ambient manifold, which we take to be Euclidean $\mathbb{R}^n$, such that

$$(1) \qquad \frac{\partial}{\partial t} F = \mathbf{H},$$

where $\mathbf{H}$ is the mean curvature vector on $M_t = F(t, M)$. By (1), the volume of M_t
decreases according to

$$(2) \qquad \frac{d}{dt}(vol(M_t)) = - \int_{M_t} H^2 \, dvol;$$

in fact, (1) is the L^2-gradient flow for the volume.

In codimension 1, that is, if $n = m + 1$, which we will assume from now on
unless otherwise stated, the mean curvature vector is given by

$$\mathbf{H} = -H\nu,$$

where ν is a smooth local unit normal vector field on $M_t \subset \mathbb{R}^n$ and where $H =
\text{trace}(\nabla \nu)$ is (m times) the (scalar) mean curvature of M_t in direction ν. Finally,

---

1991 *Mathematics Subject Classification.* 35B99, 58G99.

if for each t we denote as $F^*(\mu_{\text{eucl}})$ the pull-back of the Euclidean metric on $\mathbb{R}^n$ to M via $F(t, \cdot)$ and as $\Delta_{F^*(\mu_{\text{eucl}})}$ the induced Laplace–Beltrami operator on M, then

$$\mathbf{H} = \Delta_{F^*(\mu_{\text{eucl}})} F;$$

that is, mean curvature flow is described by the (degenerate) parabolic evolution equation

(3) $$\frac{\partial}{\partial t} F = \Delta_{F^*(\mu_{\text{eucl}})} F$$

for F, with Cauchy data

(4) $$F(0, \cdot) = F_0 : M \to \mathbb{R}^n$$

a parametrization of the initial hypersurface $M_0 = F_0(M)$.

As an example, consider the motion of a sphere $M_0 = M = \partial B_{R_0}(0; \mathbb{R}^n)$ of radius R_0. By symmetry, $M_t = \partial B_{R(t)}(0; \mathbb{R}^n)$ for each t, whence (2) translates into the equation

$$\frac{\partial}{\partial t} R = -\frac{m}{R}, \quad R_{|t=0} = R_0.$$

This can easily be solved to obtain

$$R(t) = \sqrt{R_0^2 - 2mt}.$$

In particular, the motion becomes extinct after time $T = \frac{R_0^2}{2m}$, where (M_t) vanishes in a "round" point.

Similarly, it was shown by Gage [49], Gage–Hamilton [50] and Grayson [56] that any embedded curve M_0 in the plane has a smooth evolution (M_t) under mean curvature or curve-shortening flow until (M_t) vanishes in a "round" point of asymptotically circular shape. (See Grayson [57] for corresponding results for curves on surfaces.)

In higher dimensions, the same behavior was demonstrated by Huisken [66] for convex initial hypersurfaces. Ecker–Huisken [32] showed that an initial locally Lipschitz graph $M_0 = \{(x', u_0(x')) : x' \in \mathbb{R}^m\}$ evolves by mean curvature through a family of such graphs M_t for all $t \geq 0$.

The proofs are based on the maximum principle and clever curvature estimates.

There are cases, however, where the maximum principle cannot be applied and where, instead of becoming spherical and shrinking to a single point, the surfaces M_t develop singularities of a different nature.

The simplest kind of such behavior is demonstrated by an axi-symmetric dumbell which may tear apart before its lobes have shrunk to points. In this case, and, more generally, for rotationally symmetric surfaces in R^3, it has been shown by Altschuler–Angenent–Giga [3] and Soner–Souganidis [107] that the flow can be uniquely continued with the resulting pieces past the blow-up time, yielding a unique, global "weak" solution of the mean curvature flow.

Can such a unique extension of the flow always be found? A (partial) answer to this question can be obtained if we consider the problem in a broader context.

2. Harmonic map heat flow

This flow is perhaps the one which is best understood among all geometric flows so far. Let M, N be smooth, Riemannian manifolds, $\dim M = m$. For simplicity, we take M and N to be compact and without boundary. By Nash's embedding theorem we may assume that N is isometrically embedded in $\mathbb{R}^n$ for some n. For a smooth map $u = (u^1, \ldots, u^n) : M \to N$ define the energy

$$E(u) = \frac{1}{2} \int_M |\nabla u|^2 \, dvol.$$

Given such a map $u_0 : M \to N$, we then seek to find a family of maps $(u(t))_{t \geq 0}$ that dissipate the energy $E(u_0)$ most efficiently subject to the target constraint $u(t, M) \subset N$. This can be achieved by following the L^2-gradient flow of E; that is, we look for a map $u : \mathbb{R}_+ \times M \to N \subset \mathbb{R}^n$ solving the equation

$$(5) \qquad \frac{\partial}{\partial t} u - \Delta u = A(u)(\nabla u, \nabla u) \perp T_u N$$

with initial condition

$$(6) \qquad u_{|t=0} = u_0.$$

Here, $T_p N \subset T_p \mathbb{R}^n \cong \mathbb{R}^n$ denotes the tangent space to N at a point $p \in N$ and "$\perp$" means orthogonal with respect to the inner product $\langle \cdot, \cdot \rangle$ in $\mathbb{R}^n$. $A(p) : T_p N \times T_p N \to (T_p N)^\perp$ is the second fundamental form of N at p. By orthogonality $\frac{\partial}{\partial t} u \perp A(u)(\nabla u, \nabla u)$ there holds

$$(7) \qquad \frac{d}{dt} E(u(t)) = -\int_M \left| \frac{\partial}{\partial t} u \right|^2 dvol$$

analogous to (2). In particular, for any classical solution of the Cauchy problem (5), (6) and any $0 \leq s \leq t \leq \infty$ we have

$$(8) \qquad E(u(t)) \leq E(u(s)) \leq E(u_0), \qquad \int_0^\infty \int_M \left| \frac{\partial}{\partial t} u \right|^2 dvol \, dt \leq E(u_0).$$

Of particular interest are time-independent, "stationary" solutions of (5), or "harmonic maps". Harmonic maps generalize the concept of harmonic function to mappings between Riemannian manifolds and therefore are of fundamental importance in geometry; see Eells–Lemaire [33], [34] for a survey. An encyclopedic list of references on harmonic maps and their applications was assembled by Burstall–Lemaire–Rawnsley [18]. If $M = S^1$, a harmonic map simply corresponds to a closed geodesic in N.

The following results are known concerning existence, uniqueness, and regularity of solutions to the evolution problem (5), (6).

2.1. Classical solutions. If K_N, the sectional curvature of N, is non-positive, then for any smooth map $u_0 : M \to N$ there exists a unique global smooth solution $u : \mathbb{R}_+ \times M \to N$ to (5), (6) which, as $t \to \infty$, converges smoothly to a harmonic map $u_\infty : M \to N$ freely homotopic to u_0 (Eells–Sampson [35]). The condition $K_N \leq 0$ can be relaxed to the requirement that the range of u_0 lies in a strictly convex geodesic ball in N (Jost [74], von Wahl [113]).

As in the case of the existence results of classical solutions for the mean curvature flow, the proofs of the above existence results rely fundamentally on the maximum principle. Moreover, the Bochner identity, that is, a differential equation for the evolution of the energy density $e(u) = \frac{1}{2}|\nabla u|^2$ is used.

2.2. Weak solutions. For general (compact) targets and initial maps these results in general cease to hold true. In fact, there is no harmonic map of degree $+1$ from T^2 to S^2 (Eells–Wood [36]); hence the flow (5) starting from an initial map $u_0 : T^2 \to S^2$ of degree $+1$ cannot have a global smooth solution $u : \mathbb{R}_+ \times T^2 \to S^2$ that converges smoothly, as $t \to \infty$, to a harmonic map. On the other hand, it is still possible to obtain global "weak" solutions in the "energy class" defined by relation (8).

That is, instead of classical solutions and smooth evolutions in the space of smooth maps $u : M \to N$ we consider maps in the Sobolev space

$$H^{1,2}(M;N) = \{u \in H^{1,2}(M;\mathbb{R}^n) : u(x) \in N \text{ almost everywhere}\};$$

in other words, measurable maps $u : M \to N$ with finite energy. Note that, for a flow $u : \mathbb{R}_+ \times M \to N$ satisfying $E(u(t)) \leq E(u_0)$ for almost every t, equation (5) can be interpreted in the distribution sense.

However, as regards uniqueness, a subtle dependence on the dimension of the domain is observed. For reasons that will be clear in a moment, we can distinguish "sub-critical" and "super-critical" cases, as well as a border-line " critical" case.

2.2.1. *The sub-critical case.* This is the case $m = 1$; that is, $M = S^1$. In this case, in view of the energy estimate (8) and Sobolev's embedding $H^{1,2}(S^1;N) \hookrightarrow C^{0,1/2}(S^1;N)$ we have uniform Hölder estimates for any smooth solution u to (5), (6), uniformly in time. Standard results in the regularity theory for quasi-linear parabolic systems, including (5), then give smooth a-priori bounds for u, whence u can be extended as a global classical solution to (5). (See for instance Ladyženskaya–Solonnikov–Ural'ceva [80] for the required regularity results.)

2.2.2. *The critical case.* If $m = 2$, we have the following result (Struwe [108]).

For any data $u_0 \in H^{1,2}(M;N)$ there exists a unique global weak solution $u : \mathbb{R}_+ \times M \to N$ in the energy class (8), satisfying equation (5) in the distribution sense. The solution is smooth away from finitely many singular points $(\bar{t}_i, \bar{x}_i)$, $1 \leq i \leq I$, where $I \leq CE(u_0)$. As $t \to \infty$ suitably, $u(t)$ converges weakly in $H^{1,2}(M;N)$ to a (smooth) harmonic map $u_\infty : M \to N$. At each singularity $(\bar{t}, \bar{x})$ a non-constant harmonic sphere separates in the sense that for suitable sequences $x_k \to \bar{x}$, $t_k \nearrow \bar{t}$, $R_k \searrow 0$ the rescaled maps

$$u_k(x) = u(t_k, \exp_{x_k}(R_k x)) \to \tilde{u} \quad \text{in} \quad H^{2,2}_{\text{loc}}(\mathbb{R}^2;N),$$

where $\tilde{u} : \mathbb{R}^2 \to N$ is a non-constant smooth harmonic map with finite energy $E(\tilde{u}) \leq E(u_0)$ that can be lifted conformally and extended smoothly to a non-constant harmonic map $\bar{u} : S^2 \to N$.

In Struwe [108] uniqueness was only established in the class of almost everywhere smooth solutions, satisfying (8). Uniqueness in the energy class, that is, only assuming (8), was subsequently demonstrated by Rivière [88] and Freire [44], [45]. For some time the question whether singularities actually may occur or whether the flow (5) for general targets may simply fail to converge smoothly as $t \to \infty$ remained open. The result of Chang–Ding–Ye [19] finally showed that singularities, indeed, may develop in finite time from smooth data and that the above result in this regard cannot be improved.

A special property of $m = 2$ is the fact that in this dimension the energy E is *conformally invariant*; in fact, rescaling the domain coordinates is used repeatedly as a tool in the proofs and it is crucial that the energy norm is conserved in the process.

2.2.3. *The super-critical case.* For $m \geq 3$ the problem (5), (6) still admits global weak solutions which are smooth away from a "small" singular set Σ (Chen–Struwe [**22**]). However, these solutions no longer are unique in the energy class (Coron [**25**]). Existence and partial regularity strongly rely on a "monotonicity formula" from Struwe [**110**] for a scale-invariant, weighted energy. Analogous monotonicity formulas for geometric flows are due to Giga–Kohn [**53**] and, for the mean curvature flow, to Huisken [**67**]. They prove to be a powerful tool also for the analysis of singularities. More recently Hamilton [**61**] has systematically investigated monotonicity formulas also for other geometric flows. Monotonicity formulas originate in the study of minimal hypersurfaces and in the analysis of nonlinear elliptic problems.

However, an example of Angenent–Ilmanen–Velazquez [**10**] shows that a monotonicity estimate is not sufficient for showing uniqueness of weak solutions to (5), (6).

2.3. Regularity results for harmonic maps. The results for the evolution problem (5) are mirrored in the known regularity results for solutions of the time-independent problem, that is, harmonic maps.

2.3.1. *Classical solutions.* For non-positively curved targets or under appropriate smallness assumptions on the range, regularity results, a priori estimates and comparison principles for harmonic maps again can be obtained by clever application of the maximum principle (Hildebrandt–Kaul–Widman [**64**], Jäger–Kaul [**71**], Hildebrandt–Jost–Widman [**65**])

2.3.2. *Regularity of weak solutions.* Even without any curvature restrictions, weakly harmonic maps, that is, mapping $u \in H^{1,2}(M;N)$ satisfying the equation

$$(9) \qquad\qquad -\Delta u = A(u)(\nabla u, \nabla u)$$

in the distribution sense, are smooth in the sub-critical case $m = 1$. By a beautiful result of Hélein [**63**], the same is true in the critical dimension $m = 2$. However, due to conformal invariance, one no longer can obtain smooth a priori estimates, for example, for harmonic maps $u : S^2 \to N$ in terms of the energy. The best one can hope for is compactness of families of harmonic maps with bounded energy modulo separation ("bubbling off") of harmonic spheres (Sacks–Uhlenbeck [**90**]; see also Wente [**114**] for a related problem). Bethuel [**12**], Freire–Müller–Struwe [**47**] obtain similar weak compactness results for "Palais–Smale sequences". The proof of Hélein makes use of the fact that, in a suitable frame for the pull-back bundle $u^{-1}TN$, equation (9) exhibits a determinant structure. (See also Section 6.4.2 below.)

In the super-critical case $m \geq 3$, by contrast, in general weakly harmonic maps will be singular. In fact, Rivière [**89**] has constructed an example of a nowhere continuous weakly harmonic map. Moreover, even energy-minimizing (and hence weakly harmonic) maps may be discontinuous, as the example of the map $u(x) = x/|x| : B_1(0;\mathbb{R}^m) \to S^{m-1} \subset \mathbb{R}^m$ shows. (Brezis–Coron–Lieb [**16**], Lin [**81**]). However, partial regularity results for minimizers are available (Schoen–Uhlenbeck [**95**], [**96**], Giaquinta–Giusti [**51**], Simon [**104**], [**105**]), based on a monotonicity formula for a scale-invariant weighted energy. Similar, slightly weaker results hold for weakly harmonic maps which, in addition, are "stationary" points of the energy with respect to variations of parameters in the domain (Evans [**37**]).

Thus, from the point of view of regularity theory, the previous division is repeated, in particular the conformal case $m = 2$ is distinguished as a well-behaved limit case.

3. Mean curvature flow

We now return to mean curvature flow and try to see if, depending on the dimension, a similar pattern of sub- or super-critical behavior can be observed in mean curvature flow after the break-down of regularity. In particular, we seek to identify a critical dimension with phenomena analogous to Section 2.2.2. First we need to extend the classical mean curvature flow beyond the first singular time.

3.1. Global weak solutions. In fact, there are many possible such extensions.

3.1.1. *The level set flow.* The level set flow of Evans–Spruck [**39**], [**40**], [**41**], [**42**] and Chen–Giga–Goto [**21**], building on ideas of Osher–Sethian [**85**], exchanges the Lagrangean with the Eulerian view point. That is, we no longer follow an evolving family of hypersurfaces $M_t = F(t, M)$ through space; instead, M_t is represented as level set

$$M_t = \{x \in \mathbb{R}^n : u(t, x) = 0\}$$

of some function $u : \mathbb{R}_+ \times \mathbb{R}^n \to \mathbb{R}$, evolving in a fixed coordinate system on $\mathbb{R}^n$. Equation (1) then translates into the (degenerate) parabolic equation

$$(10) \qquad \partial_t u = |\nabla u| \operatorname{div}\left(\frac{\nabla u}{|\nabla u|}\right) = \Delta u - \frac{\partial_i u \partial_i \partial_j u \partial_j u}{|\nabla u|^2}$$

for u. Here, $\partial_i = \frac{\partial}{\partial x_j}$; moreover, we tacitly sum over $1 \le i, j \le n$. In fact, if u satisfies (10), *all* its level sets evolve by mean curvature.

Employing the notion of viscosity solution defined by Crandall–Lions [**26**], it is possible to show that the Cauchy problem for (10) with Cauchy data

$$(11) \qquad\qquad u_{|t=0} = u_0 \in C^0(\mathbb{R}^n)$$

satisfying the condition $u_0(x) \equiv const$ for $|x| \ge R_0$ (corresponding to the assumption that all but one initial level surfaces are compact) admits a unique global weak solution $u \in C^0(\mathbb{R}_+ \times \mathbb{R}^n)$, and $u(t, x) \equiv const$ for $|x|^2 + 2mt \ge R_0^2$.

The motion of any level set

$$\Gamma_t = \{x \in \mathbb{R}^n : u(t, x) = 0\}$$

coincides with its classical evolution under mean curvature flow as long as Γ_t is smooth.

Moreover, comparison principles are valid and the level set flow is "natural" in the sense that it commutes with continuous reparametrizations $u \mapsto \Phi \circ u$. However, a serious problem for the geometric interpretation of the flow defined by a solution u of (10) arises due to possible "fattening" of the level set Γ_t in the sense that Γ_t has non-empty interior. In $\mathbb{R}^3$, such fattening for $t > 0$ will occur, for instance, if, in a neighbourhood of the origin, Γ_0 contains the coordinate axes. In order to obtain geometrically reasonable weak solutions for (1) from Γ_t in this case one might be tempted to (arbitrarily) define

$$M_t = \partial \Gamma_t \quad \text{for} \quad t > 0$$

and to think of Γ_0 as being covered twice.

3.1.2. *Brakke flows.* While the latter is unacceptable for classical hypersurfaces, the notion of varifolds provides an elegant and rigorous way of dealing with hypersurfaces of multiplicity different from 1. Indeed, Brakke [**14**] already in 1978 extended classical mean curvature flow to varifolds and obtained global weak solutions and partial regularity in this context, but again no uniqueness. Brakke flows and the level set flow were merged by Ilmanen [**69**], using elliptic regularization. Moreover, Ilmanen was able to complete Brakke's regularity theory and to show almost everywhere regularity for generic data.

3.1.3. *Approximation via phase field equations.* Finally, mean curvature flow arises in the singular limit $\varepsilon \to 0$ of certain semi-linear heat equations, for instance, the equation

$$(12) \qquad\qquad \partial_t u - \Delta u + \varepsilon^{-2} V(u) = 0,$$

where $V(u) = (1 - |u|^2)^2$ is a two-well potential, modelling the motion of phase boundaries (Allen–Cahn [**2**], de Mottoni–Schatzman [**28**], Bronsard–Kohn [**17**], Chen [**20**], Evans–Soner–Souganidis [**38**], Ilmanen [**68**], Soner [**106**]).

For $\varepsilon > 0$, the Cauchy problem for equation (12) can be solved uniquely for all time. By work of Ilmanen [**68**], as $\varepsilon \to 0$ the 0-level sets of the corresponding solutions u_ε (for suitable initial data $u_{\varepsilon 0}$) converge to a Brakke flow. However, as was also remarked by Ilmanen [**68**], any of the (non-unique) Brakke flows can be obtained in this way (and all of these are contained in the level set flow).

The situation therefore is analogous to the situation encountered in the harmonic map heat flow (5) in the super-critical case $m \geq 3$. In this regard it is highly interesting to observe that also the harmonic map heat flow may be obtained in the limit $\varepsilon \to 0$ from equations like (12); conversely, via its interpretation as a singular limit of phase field equations, mean curvature flow can be considered as renormalized harmonic map heat flow with infinite energy (Ilmanen [**68**]).

Phase field equations and systems of such equations also offer a way of approximating mean curvature flow in codimension $n - m > 1$ and of defining a concept of weak solution in this case; see De Giorgi [**27**], Ambrosio–Soner [**4**], Jerrard–Soner [**72**].

3.2. Is mean curvature flow always super-critical?

3.2. Is mean curvature flow always super-critical? As we have seen, none of the above notions of generalized mean curvature flow seems capable of dealing with the problem of non-uniqueness after a first geometric singularity. Is there a better notion?

In view of the above examples it seems reasonable to restrict our focus to parametrized (immersed) hypersurfaces. By analogy with the results from Section 2, uniqueness of weak solutions should hold in "critical" cases where some conformally invariant "energy" is non-increasing under the flow. When do such energy functionals exist and how are they defined? Clearly the volume is not dilation invariant. Only curvature integrals seem to be reasonable candidates. In which dimensions can we hope to find flow-invariant bounds for them? Recent examples of Angenent–Ilmanen–Velazquez [**10**] show that (in contrast to the case $m = 2$) for $m = 3, \dots, 6$ even a codimension 1 hypersurface of revolution may evolve non-uniquely after a first singularity, and similar non-uniqueness results hold for initial hypersurfaces near a minimal cone in $\mathbb{R}^8$ and in higher dimensions. This leaves only dimensions $m = 2$ and 1.

In $m = 2$ dimensions, for any family of hypersurfaces M_t, $0 \leq t < T$, smoothly evolving under mean curvature flow, the Gauss–Bonnet formula and (2) allow to express the L^2-norm of the second fundamental form A on M_t as

$$\int_{M_t} |A|^2 \, do = \int_{M_t} (k_1^2 + k_2^2) \, do = \int_{M_t} (H^2 - 2K) \, do$$

$$= -\frac{d}{dt}(area(M_t)) - 4\pi\chi(M_0),$$

where $k_{1,2}$ denote the principal curvatures on M_t, $K = k_1 k_2$ is the Gauss curvature, and where $\chi(M_0) = \chi(M_t)$ is the Euler characteristic. In particular, we have the space-time estimate

(13) $$\int_0^T \int_{M_t} |A|^2 \, do \, dt \leq area(M_0) + CT.$$

Certain regularity results based on this estimate (and local refinements) have been obtained by Ecker [**31**] and Ilmanen [**70**]. However, an example of Angenent–Chopp–Ilmanen [**9**], partly based on numerical evidence, shows that (13) in general does not give sufficient control on the limit surface M_T to ensure uniqueness of the mean curvature flow for times $t > T$. Thus it seems that the response to the question we posed at the end of Section 1 is negative and that the mean curvature flow is super-critical in any dimension $m \geq 2$.

3.3. The critical case. In fact, the only case where mean curvature flow seems to be globally well-posed is the case of immersed curves M_t in the plane. (Recall that for *embedded* plane curves the maximum principle gives unique classical solutions.) The desired dilation-invariant "energy" integral was identified by Abresch–Langer [**1**]. Indeed, if k denotes the curvature along M_t with respect to a local frame, the total absolute curvature

(14) $$\int_{M_t} |k| ds$$

is non-increasing in t. Using (14), Angenent [**5**], [**6**] shows that for any immersion $F_0 : S^1 \to \mathbb{R}^2$ the mean curvature flow (1) admits a family of piecewise immersed curves $M_t = F(t, S^1)$, $t \geq 0$ as a global weak solution. The curves M_t are smooth for all but finitely many times where curvature concentrates along shorter and shorter subarcs of M_t. If the initial curve is convex, that is, if its curvature does not change sign, a condition which is preserved under the flow, then by blowing up the singularities via rescaling, one either finds a self-similar solution of the mean curvature flow (previously studied and classified by Abresch–Langer) or a translating solution, Grayson's "grim reaper" (Angenent [**8**]). Finally, it seems that the flow $(M_t)_{t \geq 0}$ is unique (Angenent [**7**]); that is, there is complete analogy to the results of Section 2.2.2.

4. Further evidence

Other geometric flows provide further evidence for a universal pattern of sub- and super-critical behavior, divided by a borderline critical dimension which is distinguished by conformal invariance of the energy and uniqueness of weak solutions in the energy class. We cite two more examples.

4.1. The Yamabe flow. In 1991, Ye [**115**] gave a new proof of the Yamabe conjecture (which had previously been solved by Aubin [**11**] and Schoen [**94**]) in the—apparently—difficult case of a locally conformally flat manifold by solving the related heat equation, the Yamabe flow. Note that by their intrinsic meaning, the Yamabe energy functional and the Yamabe flow are conformally invariant. In the case of a locally conformally flat m-manifold then, the Yamabe flow can be lifted to the m-sphere via the developing map of Schoen–Yau [**97**]. An ingenious argument, based on ideas from Gidas–Ni–Nirenberg [**52**], then combines conformal invariance, Alexandrov reflection and the maximum principle to obtain a uniform Harnack estimate for the flow and hence its asymptotic convergence to a solution of the Yamabe problem.

The Yamabe flow is related to Richard Hamilton's Ricci flow; see Hamilton [**62**].

4.2. Yang–Mills heat flow. For a connection $d_A = d_0 + A$ on a vector bundle X over an m-dimensional base manifold M with fibres isomorphic to $\mathbb{R}^n$ and structure group $G \subset SO(n)$, with A a 1-form with values in the Lie algebra of G, the Yang–Mills functional is

$$YM(A) = \int_M |F|^2 \, dvol,$$

where $F = d_A \circ d_A$ is the curvature of d_A. The Yang–Mills functional is invariant under gauge transformations; moreover, it is conformally invariant in dimension $m = 4$. For fixed background connection d_0, the Yang–Mills heat flow is a degenerate parabolic equation for the connection 1-form A. The flow becomes strongly parabolic if we impose, for instance, the Coulomb gauge condition $d_A^* A = 0$.

By results of Råde [**87**], the Cauchy problem for the Yang–Mills heat flow in the subcritical dimensions $m = 2$ or 3 admits a unique global solution which converges as $t \to \infty$ to a Yang–Mills connection on X.

In the critical dimension $m = 4$, similar to the behavior of (5) in case $K_N \leq 0$, the same results hold if the underlying bundle is holomorphic and stable (Donaldson–Kronheimer [**29**]). Without any additional differential topological conditions the exact analogue of the results in Section 2.2.2 for the conformal harmonic map heat flow is observed (Struwe [**111**], Schlatter [**92**]). However, in contrast to the harmonic map heat flow, so far no examples of finite-time blow-up have been found. In fact, there is evidence that blow-up may never occur in finite time (Schlatter–Struwe–Tahvildar-Zadeh [**93**]). (In dimension $m > 4$ not even the existence of weak solutions is known.)

5. Relevance in mathematical physics

Some of the above geometric flow equations may be of significance in physics to describe evolution phenomena where diffusion is the dominant driving mechanism and momentum or inertia can be ignored. However, as regards the Yang–Mills or Ricci (that is, the Einstein) equations, "real" physics requests a $(1 + 3)$-dimensional *Lorentzian* space-time and the corresponding physical models give rise to systems of *hyperbolic* partial differential equations. Does the above pattern also manifest itself for this class of evolution problems and to what extent? In particular, if the associated energy is conformally invariant, are the evolution problems of mathematical physics well-posed in the energy class?

To guide us, we consider the Cauchy problem for three model equations of quite diverse structure: scalar field equations, Yang–Mills equations and wave maps. It is natural to consider initial data in Sobolev spaces. For $s \geq 0$, $H^s(\mathbb{R}^m) = H^{s,2}(\mathbb{R}^m)$ is the space of functions $u \in L^2(\mathbb{R}^m)$ with square integrable distributional derivatives up to order s. As is well-known, the Cauchy problem for the linear wave equation

(15) $$\Box u = \partial_t^2 - \Delta u = 0 \quad \text{on} \quad \mathbb{R} \times \mathbb{R}^m$$

is well-posed for initial data

(16) $$u_{|t=0} = u_0, \partial_t u_{|t=0} = u_1 \in H^s \times H^{s-1}(\mathbb{R}^m),$$

for any $s \in \mathbb{R}$, and there is exact dependence (no "loss of derivatives") for the solution on the data in these spaces. Finally, to facilitate notation, we let $z = (t, x) = (x^\alpha)_{0 \leq \alpha \leq m}$ denote space-time coordinates on $\mathbb{R} \times \mathbb{R}^m$ and $D = (\partial_t, \nabla)$.

5.1. Nonlinear scalar field equations.
In the 50's semi-linear wave equations

(17) $$\Box u + f(u) = 0 \qquad \mathbb{R} \times \mathbb{R}^m$$

were introduced in relativistic particle physics, where $f = F'$ with a smooth, non-negative potential F, typically of the form

$$F(u) = \frac{|u|^p}{p} + \text{lower order terms}$$

(Schiff [**91**]). As model problem, for $p > 2$ we consider the equation

(18) $$\Box u + u|u|^{p-2} = 0 \quad \text{on} \quad \mathbb{R} \times \mathbb{R}^m$$

with Cauchy data (16). Equation (17) implies the conservation law

(19) $$\partial_t e(u) - \text{div } m(u) = 0,$$

where

$$e(u) = \frac{1}{2}|Du|^2 + F(u)$$

denotes the energy density and

$$m(u) = \nabla u \partial_t u.$$

Integrating (19) over a truncated light cone

$$K_S^T(z_0) = \{(t, x) : S \leq t \leq T, |x - x_0| \leq t_0 - t\}$$

through a point $z_0 = (t_0, x_0)$, for all $S \leq T \leq t_0$ we obtain the inequality

(20) $$E(u; D(T; z_0)) \leq E(u; D(S; z_0))$$

for the local energy

$$E(u; D(t; z_0)) = \int_{D(t;z_0)} e(u) \, dx$$

on the slice $D(t; z_0) = K_S^T(z_0) \cap \{t\} \times \mathbb{R}^m$.

In particular, information propagates with speed at most 1; moreover, for any smooth solution u of (17) with spatially compact support the total energy

$$E(u(t)) = \int_{\mathbb{R}^m} e(u(t)) \, dx$$

is conserved. Using this fact, global weak solutions, for instance, to our model equation (18), can be constructed for any $p \geq 2$ and any $m \in \mathbb{N}$ (Segal [98], Lions [82]). Finally, note that if u solves (18) so does

$$u_R(t, x) = R^{\frac{2}{p-2}} u(Rt, Rx)$$

for any $R > 0$ and

$$E(u_R(t)) = R^{\frac{2p}{p-2} - m} E(u(Rt)).$$

That is, E scales invariantly if and only if $p = \frac{2m}{m-2} = 2^*$. Thus, $p = 2^*$ is the critical growth exponent for equation (17). Observe that 2^* also appears as limiting exponent for Sobolev's embedding $H^1(\mathbb{R}^m) \hookrightarrow L^p(\mathbb{R}^m), 2 \leq p \leq 2^*$.

5.1.1. *The sub-critical case.* If $p < 2^*$, by results of Jörgens [73], Pecher [86], Brenner–von Wahl [15] the Cauchy problem for equation (17) admits global smooth solutions for smooth data in dimensions $m \leq 9$. Weak solutions are unique in the "energy-class" defined by relation (20) in all dimensions (Ginibre–Velo [54]).

5.1.2. *The critical case.* Regularity for large data in $m = 3$ space dimensions for the critical power $p = 6$ was demonstrated by Struwe [110] for radial solutions; the symmetry condition was removed by Grillakis [58]. In higher dimensions corresponding regularity results were subsequently obtained by Grillakis [59], Shatah–Struwe [100], Ginibre–Velo [55], for $m \leq 9$. Moreover, in *any* dimension $m \geq 3$, for data $(u_0, u_1) \in H^1 \times L^2$, that is, of finite energy, the Cauchy problem (18), (16) admits a unique global weak solution in a class which is closely related to the energy-class (Shatah–Struwe [100], [101]; see also Kapitanski [75]).

5.1.3. *The super-critical case.* For super-critical powers and large data no results concerning existence of global classical solutions or uniqueness of (suitable) weak solutions are known. Thus, although many of the difficulties that prevent us from proving global regularity and well-posedness of (17), (16) appear to be of only technical nature, it appears that also in the hyperbolic context scaling invariance of the energy marks the borderline between sub-critical and critical behavior, as characterized by well-posedness in the energy space, and super-critical behavior, possibly related to non-uniqueness and break-down of regularity.

5.2. Yang–Mills equations. The equation for a Yang–Mills connection $d_A = d + A$ on Minkowski space, where $A = A_\alpha dx^\alpha = A_0 \, dt + A_i \, dx^i$, take the form

$$(21) \qquad 0 = d_A^* F = d^* dA + A \# DA + A \# A \# A \quad \text{on} \quad \mathbb{R} \times \mathbb{R}^m$$

with certain bi-linear and cubic expressions involving A and its space-time derivatives. Here, d is the ordinary exterior differential, acting componentwise on sections, and d_A^* is the adjoint of d_A in the Minkowski metric. $F = d_A \circ d_A = dA + \frac{1}{2}[A, A]$ is the curvature of d_A. Its L^2-norm, the Yang–Mills energy

$$YM(A(t)) = \int_{\{t\} \times \mathbb{R}^m} |F|^2 dx$$

is conserved by any classical solution of (21). Note that (21) is gauge-invariant. Imposing Lorentz gauge $d^* A = 0$, we have

$$d^* dA = (d^* d + dd^*)A = \Box A = (\Box A_\alpha) \, dx^\alpha;$$

that is, (21) is a system of quasi-linear hyperbolic partial differential equations which differs in structure from (17) only by a term $A \# DA$ with linear dependence on DA. However, the temporal gauge $A_0 = 0$ or the Coulomb gauge $\partial_i A_i = 0$ turn out to be more useful.

We consider admissible (gauge-compatible) initial data

$$(22) \qquad (A_{|t=0}, F_A \,|_{t=0}) \in H^s \times H^{s-1}(\mathbb{R}^m).$$

In the case of interest in physics, that is, $m = 3$, by results of Eardley–Moncrief [**30**], for $s = 2$ the Cauchy problem (21), (22) admits a global, unique solution of class H^2 in the temporal gauge. Recently, Klainerman–Machedon [**78**] have sharpened this result considerably, by showing that the solutions of Eardley–Moncrief continuously depend on the data in the space $H^1 \times L^2$; that is, the initial value problem (21), (22) for $m = 3$ is well-posed (in the temporal gauge) in the energy space. The proofs of Klainerman–Machedon draw on sharp space-time estimates for "null-forms" (Klainerman–Machedon [**77**]) that seem to exploit the structure hidden in the nonlinear terms in equation (21) in an optimal way that hardly leaves any room to extend these results to $m \geq 4$. On the other hand, by comparison with semi-linear wave equations (18) the case $m = 3$ for equation (21) should still be regarded as sub-critical. Does the appearance of a nonlinear term involving DA therefore give rise to a shift in the pattern we observed so far with onset of super-critical behavior at or below the conformal dimension? To answer this question, research on equation (21) also in the non-physical space dimension $m = 4$ will be needed.

Here, we will go even one step further and consider quasilinear hyperbolic systems with quadratic dependence in the space-time gradient.

6. Wave maps

As a final example, we consider "wave maps" $u : \mathbb{R} \times \mathbb{R}^m \to N \subset \mathbb{R}^n$, satisfying the equation

$$(23) \qquad \Box u = (\partial_t^2 - \Delta)u = A(u)(Du, Du) \perp T_u N,$$

where, as in Section 2, N is a compact Riemannian manifold without boundary isometrically embedded in $\mathbb{R}^n$, and where $A(p) : T_p N \times T_p N \to (T_p N)^\perp$ denotes the second fundamental form of $N \subset \mathbb{R}^n$. That is, wave maps are harmonic maps of Minkowski space M^{1+m} into N. Note that $\partial_t u \in T_u N$. Hence (23) implies the conservation law

$$(24) \qquad 0 = \langle \Box u, \partial_t u \rangle = \frac{d}{dt} e(u) - \mathrm{div}(\langle \nabla u, \partial_t u \rangle)$$

for the energy density $e(u) = \frac{1}{2}|Du|^2$, yielding the analogue of (20) and, in particular, that

$$(25) \qquad E(u(t)) = \int_{\{t\} \times \mathbb{R}^m} e(u) \, dx \equiv E(u(0))$$

for any smooth solution of (23) with Du of spatially compact support. Moreover, information propagates with speed at most 1.

If $N = S^{n-1} \subset \mathbb{R}^n$, equation (23) becomes

$$(26) \qquad \Box u = \left(|\nabla u|^2 - |\partial_t u|^2\right) u.$$

A related problem thus is given by the scalar equation

$$(27) \qquad \Box u = |\nabla u|^2 - |\partial_t u|^2.$$

This equation can be solved explicitly via the transformation $v = e^u$ which transforms (27) into the linear wave equation

$$(28) \qquad \qquad \Box v = 0.$$

However, (27) lacks the geometric interpretation of (23). (In fact, (27) can be interpreted as the equation for wave maps to the real line, equipped with a *singular* metric (Klainermann [76]).)

Again we are interested in local and global well-posedness of the Cauchy problem for (23) with suitably regular data

$$(29) \qquad \qquad (u_{|t=0} = u_0, \partial_t u_{|t=0} = u_1) : \mathbb{R}^m \to TN;$$

that is, $u_1(x) \in T_{u_0(x)}N$ for all x. We consider data $(u_0, u_1) \in H^s \times H^{s-1}$ ("of class H^s"). Remark that by (25) the energy class corresponds to $s = 1$. As in the parabolic setting considered above, E is dilation invariant in $m = 2$ dimensions. Thus, $m = 2, s = 1$ is the critical, conformal case.

6.1. Local classical solutions. For $s = 2$ and $m \le 3$ the Cauchy problem (23), (29) has a local unique solution of class H^s (Klainerman–Machedon [77], Struwe [112]). In fact, at least in low dimensions, the initial value problem is well-posed in H^s for $s > \frac{m}{2}$, as is consistent with (27), (28) (Klainerman–Machedon [79]). If $m = 1$, the initial value problem is globally well-posed in H^2 (Shatah [99]); in fact, even in H^1 (Yi Zhou [117]). In all cases, the solutions also preserve higher regularity of the data.

6.2. Global weak solutions. For symmetric target spaces, equation (23) can be rewritten in the form of a system of conservation laws. For instance, letting "$\wedge$" denote the exterior product in $\mathbb{R}^3$, the equation (26) for wave maps into S^2 is equivalent to the system

$$(30) \qquad \begin{aligned} &0 = \Box u \wedge u = \partial_t(\partial_t u \wedge u) - \operatorname{div}(\nabla u \wedge u), \\ &0 = \Box |u|^2. \end{aligned}$$

The energy bound (25) then is sufficient to pass to the limit in a suitable sequence of approximate solutions (Shatah [99]). Shatah's result was generalized to classical Lie groups and their quotients by Freire [46]. Finally, for $m = 2$ and small initial energy, Yi Zhou [116] obtained global weak solutions for the Cauchy problem for wave maps into any compact symmetric space.

6.3. Singularities, non-uniqueness. In dimensions $m \ge 3$ solutions to (23) may become singular in finite time. In fact, if $m = 3$ Shatah [99] has constructed non-constant self-similar solutions to (15) of the form

$$u(t,x) = v\left(\frac{x}{t}\right)$$

where $v : \mathbb{R}^3 \to S^3$ is a smooth map with range (essentially) a hemisphere. If $m \ge 4$, examples of Shatah–Tahvildar-Zadeh [103] show that self-similar blow-up is possible even when the range is convex. Moreover, if $m \ge 3$ the initial value problem for (23) with data of class H^1_{loc} in general cannot be solved uniquely in the energy class (Shatah–Tahvildar-Zadeh [103]).

6.4. The conformal case. By contrast, if $m = 2$ there are no non-trivial self-similar solutions of (23) (Struwe [**112**]). Moreover, under additional symmetry conditions the Cauchy problem (23), (29) is globally well-posed and the solutions propagate the regularity of the data. In particular, this is the case for radial solutions $u(t, x) = u(t, |x|)$ of (23) (Christodoulou–Tahvildar-Zadeh [**23**]). Moreover, this is the case for co-rotational maps into (strongly) non-compact surfaces of revolution, defined as follows.

6.4.1. *Regularity of co-rotational wave maps.* In polar coordinates (r, ϕ) on $\mathbb{R}^2$ and (ρ, θ) on N, a co-rotational map $u : \mathbb{R} \times \mathbb{R}^2 \to N$ is given by

$$\rho = h(t, r), \theta = \phi$$

with a scalar, radial function $h : \mathbb{R} \times \mathbb{R}^2 \to \mathbb{R}$. Denoting the metric on N as

$$ds^2 = d\rho^2 + g(\rho)^2 d\theta^2,$$

then u solves (23) if and only if h solves the singular semilinear wave equation

(31) $$\Box h + \frac{f(h)}{r^2} = \partial_t^2 h - \partial_r^2 h - \frac{1}{r}\partial_r h + \frac{f(h)}{r^2} = 0,$$

where $f(\rho) = g(\rho)g'(\rho)$. Under the natural assumptions

(32) $$g(\rho) = -g(-\rho), g'(0) = 1, g(\rho) > 0 \text{ for } \rho > 0,$$

and assuming N to be strongly non-compact in the sense that

(33) $$\int_0^\infty g(\rho)\, d\rho = \infty,$$

the initial value problem for (31) with smooth radially symmetric data admits a unique global smooth solution if $g' > 0$, that is, if N is convex (Shatah–Tahvildar-Zadeh [**102**]), or, more generally, if $g(\rho) + \rho g'(\rho) > 0$ for all $\rho > 0$ (Grillakis [**60**]).

It is conjectured that conditions (32), (33) are sufficient for regularity; moreover, in case of a compact target surface, singularities should be related to non-constant harmonic spheres separating, as in Section 2.2.2 for the heat flow of harmonic maps. Finally, remark that by substituting $\varphi(t, r) = r^{-1}h(t, r)$, equation (31) is transformed into

$$\partial_t^2 \varphi - \partial_r^2 \varphi - \frac{3}{r}\partial_r \varphi + K(r\varphi)\varphi^3 = 0;$$

that is, formally φ is a radial solution of a semi-linear wave equation

$$\Box \varphi + K(r\varphi)\varphi^3 = 0 \quad \text{on} \quad \mathbb{R} \times \mathbb{R}^4$$

with a smooth bounded function K. The results in Section 5.1 thus seem to indicate that the Cauchy problem for co-rotational wave maps under assumptions (32), (33) should be well-posed in the energy space.

6.4.2. *General targets.* For general (compact) targets and maps global well-posedness of the Cauchy-problem for (23) still remains open. However, some progress towards a complete resolution of this question recently has been made by Freire–Müller–Struwe [**47**], [**48**] in showing that the set of weak solutions of (23) with bounded energy and satisfying the local energy inequality (20) (a weak form of the conservation law (24)) is weakly compact. Combining this compactness result with the approximation method of Yi Zhou [**116**] or (alternatively) with a spatially discrete approximation, Müller–Struwe [**83, 84**] then obtain global weak solutions to (23), (29) for finite energy data.

The proofs involve an equivalent reformulation of (23) for maps $u_k : \mathbb{R} \times \mathbb{R}^2 \to N$, $k \in \mathbb{N}$, as a first order Hodge system, using a parallelization of the pull-back tangent bundle $u_k^{-1} TN$, as in Hélein [63] or Christodoulou–Tahvildar-Zadeh [23]. (If TN is not already parallelizable, one first has to embed N as a totally geodesic submanifold in another compact manifold $\tilde{N}$ whose tangent bundle *is* parallelizable.) Imposing an *elliptic* space-time Coulomb gauge condition on the choice of frames, the Hodge system exhibits a determinant structure and the Hardy space estimates of Coifman–Lions–Meyer–Semmes [24] can be applied. On the other hand, the uniform energy bound (25) gives an estimate for u and related terms in the space BMO which by results of Fefferman–Stein [43] is dual to the Hardy space $\mathcal{H}^1$. Finally, by carefully analyzing the concentration set of the measures $|Du_k|^2 dz$, it is possible to pass to the distributional limit $k \to \infty$ in the Hodge systems for u_k. Similar arguments are used by Evans [37] and Bethuel [13] in their proofs of partial regularity of stationary harmonic maps.

7. Perspective

We conclude from these examples that, indeed, in many evolution problems in geometry and mathematical physics a universal pattern of sub- and super-critical behavior manifests itself, irrespective of the analytic type (parabolic or hyperbolic) of the associated flow equations. While a lot of research, in particular, on evolution problems of hyperbolic type still needs to be carried out, already it seems that invariably the criterion of scaling invariance of the natural energy associated with the flow marks the borderline between the sub- and super-critical regimes and that the initial value problem seems to be well-posed in the energy space up to and including the critical dimension.

Many of the problems of interest in mathematical physics, for instance, the Einstein equations for a 4-dimensional Lorentzian space-time, from this point of view, however, seem super-critical. Further research will be necessary to see if in these cases natural conditions, possibly of "entropy"-type, can be identified that allow to distinguish a particular class of weak solutions and to show that the initial value problem is well-posed in this class, at least for "generic" data. Of the numerous open problems in the field of geometric evolution equations, this question seems to be one of the most challenging.

Acknowledgements

I thank Princeton University and the Princeton University Mathematics Department for their hospitality and support. Attributions of results in this paper are made to the best of my knowledge. Necessarily, in view of the vast recent literature the list of references cannot be exhaustive or complete, and I wish to apologize for any errors or omissions.

References

[1] Abresch, U. – Langer, J.: *The normalized curve shortening flow and homothetic solutions.* J. Diff. Geom. **23** (1987), 175–196.

[2] Allen, S. – Cahn, J.: *A microscopic theory for antiphase boundary motion and its application to antiphase domain coarsening*, Acta Metall **27** (1979), 1084–1095.

[3] Altschuler, S. – Angenent, S. B. – Giga, Y.: *Mean curvature flow through singularities for surfaces of rotation*, J. Geom. Anal. **5** (1995), 293–358.

[4] Ambrosio, L. – Soner, H. M.: *Level set approach to mean curvature flow in arbitrary codimension*, J. Diff. Geom. **43**:4 (1996), 693–737.

[5] Angenent, S. B.: *Parabolic equations for curves on surface I*, Ann. of Math. **132** (1990), 451–483.

[6] Angenent, S. B.: *Parabolic equations for curves on surface II*, Ann. of Math. **133** (1991), 171–215.

[7] Angenent, S. B.: (private communication).

[8] Angenent, S. B.: *On the formation of singularities in the curve-shortening flow*, J. Differential Geometry **33** (1991), 601–633.

[9] Angenent, S. B. – Chopp, P. L. – Ilmanen, T.: *A computed example of nonuniqueness of mean curvature flow in $\mathbb{R}^3$*, Comm. Partial Differential Equations **20** (1995), 1937–1958.

[10] Angenent, S. B. – Ilmanen, T. – Velazquez, J. J. L.: *Nonuniqueness in geometric heat flows* (in preparation).

[11] Aubin, T.: *The scalar curvature, Differential Geometry and Relativity* (eds. Cahen and Flato), Reider, 1976.

[12] Bethuel, F.: *Weak convergence of Palais–Smale sequences for some critical functionals*, Calc. Var. **1** (1993), 267–310.

[13] Bethuel, F.: *On the singular set of stationary harmonic maps*, Manuscripta Math. **78** (1993), 417–443.

[14] Brakke, K.: *The motion of a surface by its mean curvature*. Princeton University Press, 1978.

[15] Brenner, P. – von Wahl, W.: *Global classical solutions of non-linear wave equations*, Math. Z. **176** (1981), 87-121.

[16] Brezis, H. – Coron, J.-M. – Lieb, E.: *Harmonic maps with defects*. Comm. Math. Phys. **107** (1986), 649–705.

[17] Bronsard, L. – Kohn, R. V.: *Motion by mean curvature as the singular limit of Ginzburg–Landau dynamics*, J. Differential Equations **90** (1991), 211–237.

[18] Burstall, F. E. – Lemaire, L. – Rawnsley, J.: *Harmonic maps bibliography*, 18 January 1994.

[19] Chang, K.-C. – Ding, W.-Y. -Ye, R.: *Finite-time blow-up of the heat flow of harmonic maps from surface*. J. Diff. Geom. **36** (1992), 507–515.

[20] Chen, X.: *Generation and propagation of the interface for reaction-diffusion equations*, J. Differential Equations **96** (1992), 116–141.

[21] Chen, Y.-G. – Giga, Y. – Goto, S.: *Uniqueness and existence of viscosity solutions of generalized mean curvature flow equations*. J. Diff. G. **33** (1991), 749–786 (announcement in Proc. Japan. Acad. Ser. A **65** (1985) 207–210).

[22] Chen, Y. – Struwe, M.: *Existence and partial regularity results for the heat flow for harmonic maps*. Math. Z. **201** (1989), 83–103.

[23] Christodoulou, D. – Tahvildar-Zadeh, A. S.: *On the regularity of spherically symmetric wave maps*, Comm. Pure Appl. Math. **46** (1993), 1041–1091.

[24] Coifman, R. – Lions, P.-L. – Meyer, Y. – Semmes, S.: *Compensated compactness and Hardy spaces*, J. Math. Pures Appl. **72** (1993), 247–286.

[25] Coron, J.-M.: *Nonuniqueness for the heat flow of harmonic maps*. Ann. Inst. H. Poincaré, Analyse Non Linéaire **7** (1990), 335–344.

[26] Crandall, M. G. – Lions, P.-L.: *Viscosity solutions of Hamilton–Jacobi equations*. Trans. AMS **277** (1983), 1–42.

[27] De Giorgi, E.: *Some conjectures on flow by mean curvature*, white paper (1990).

[28] de Mottoni, P. – Schatzman, M.: *Evolution geometrique d'interfaces*, C. R. Acad. Sci. Paris **309** (1989), 453–458.

[29] Donaldson, S. K. – Kronheimer, P.: *The topology of four-manifolds*, Clarendon Press, Oxford (1990).

[30] Eardley, D. – Moncrief, V.: *The global existence of Yang–Mills–Higgs fields in M^{3+1}*, Comm. Math. Phys. **83** (1982), 171–212.

[31] Ecker, K.: *On regularity for mean curvature flow of hypersurfaces*, Calc. Var. **3** (1995), 107–126.

[32] Ecker, K. – Huisken, G.: *Mean curvature evolution of entire graphs*. Annals of Math. **130** (1989), 453–471.

[33] Eells, J. – Lemaire, L.: *A report on harmonic maps*. Bull. London Math. Soc. **10** (1978), 1–68.

[34] Eells, J. – Lemaire, L.: *Another report on harmonic maps*. Bull. London Math. Soc. **20** (1988), 385–524.

[35] Eells, J. – Sampson, J. H.: *Harmonic mappings of Riemannian manifolds*. Am. J. Math. **86** (1964), 109–169.

[36] Eells, J. – Wood, J. C.: *Restrictions on harmonic maps of surfaces*. Topology **15** (1976), 263–266.

[37] Evans, L. C.: *Partial regularity for stationary harmonic maps into spheres*, Arch. Rat. Mech. Analysis, **116** (1991), 101–113.

[38] Evans, L. C. – Soner, H. M. – Souganidis, P. E.: *Phase transitions and generalized motion by mean curvature*, Comm. Pure Appl. Math. **45** (1992), 1097–1123.

[39] Evans, L. C. – Spruck, J.: *Motion of level sets by mean curvature I*. J. Diff. Geom. **33** (1991), 635–681.

[40] Evans, L. C. – Spruck, J.: *Motion of level sets by mean curvature II*. Transactions of the AMS **330** (1922), 321–332.

[41] Evans, L. C. – Spruck, J.: *Motion of level sets by mean curvature III*. J. Geom. Analysis **2** (1992), 121–150.

[42] Evans, L. C. – Spruck, J.: *Motion of level sets by mean curvature IV*. J. Geom. Analysis **5** (1995), 77–114.

[43] Fefferman, C. – Stein, E. M.: *H^p spaces of several variables*, Acta Math. **129** (1972), 137–193.

[44] Freire, A.: *Uniqueness for the harmonic map flow in two dimensions*, Calc. Var. **3** (1995), 95–105.

[45] Freire, A.: *Uniqueness for the harmonic map flow from surfaces to general targets*, Comm. Math. Helv. **70** (1995), 310–338; correction in Comm. Math. Helv. **71**:2 (1996), 330–337.

[46] Freire, A.: *Wave maps to compact Lie groups* preprint (1996).

[47] Freire, A. – Müller, S. – Struwe, M.: *Weak convergence of wave maps from (1+2)-dimensional Minkowski space to Riemannian manifolds*, Invent. Math. **130** (1997), 589–617.

[48] Freire, A. – Müller, S. – Struwe, M.: *Weak compactness of wave maps and harmonic maps*, to appear in Ann. Inst. H. Poincaré, Analyse Non Linéaire.

[49] Gage, M. E.: *Curve shortening makes convex curves circular*. Invent. Math. **76** (1984), 357–364.

[50] Gage, M. E. – Hamilton, R. S.: *The heat equation shrinking convex plane curves*. J. Diff. Geom. **23** (1986), 69–96.

[51] Giaquinta, M. – Giusti, E.: *The singular set of the minima of certain quadratic functionals*. Ann. Scuola Norm. Sup. Pisa (4) **11** (1984), 45–55.

[52] Gidas, B. – Ni, W.-M. – Nirenberg, L.: *Symmetry and related properties via the maximum principle*, Comm. Math. Phys. **68** (1979), 209–243.

[53] Giga, Y. – Kohn, R. V.: *Asymptotically self-similar blow up of semi-linear heat equations*, Comm. Pure Appl. Math. **38** (1985), 297–319.

[54] Ginibre, J. – Velo, G.: *The global Cauchy problem for the non-linear Klein–Gordon equation*, Math. Z. **189** (1985), 487–505.

[55] Ginibre, J. – Velo, G.: *Regularity of solutions of critical and subcritical nonlinear wave equations*, Nonlin. Analysis, TMA **22** (1994), 1–19.

[56] Grayson, M. A.: *The heat equation shrinks embedded plane curves to round points*. J. Diff. Geom. **26** (1987), 285–314.

[57] Grayson, M. A.: *Shortening embedded curves*. Annals of Math. **129** (1989), 71–111.

[58] Grillakis, M.: *Regularity and asymptotic behaviour of the wave equation with a critical nonlinearity*. Ann. of Math. **132** (1990), 485–509.

[59] Grillakis, M.: *Regularity for the wave equation with a critical nonlinearity*. Comm. Pure Appl. Math. **45** (1992), 749–774.

[60] Grillakis, M.: *Classical solutions for the equivariant wave map in 1 + 2 dimensions*. Preprint.

[61] Hamilton, R. S.: *Monotonicity formulas for parabolic flows on manifolds*, Comm. Anal. Geom. **1** (1993), 127–137.

[62] Hamilton, R. S.: *The formation of singularities in the Ricci flow*, International Press, Cambridge, MA, 1995.

[63] Hélein, F.: *Regularité des applications faiblement harmoniques entre une surface et une varité Riemannienne*. C. R. Acad. Sci. Paris Ser. I Math. **312** (1991), 591–596.

[64] Hildebrandt, S. – Kaul, H. – Widman, K.-O.: *An existence theorem for harmonic mappings of Riemannian manifolds*, Acta Math. **138** (1977), 1–16.

[65] Hildebrandt, S. – Jost, J. – Widman, K.-O.: *Harmonic mappings and minimal submanifolds*, Invent. Math. **62** (1980), 269–298.

[66] Huisken, G.: *Flow by mean curvature of convex surfaces into spheres*. J. Diff. Geom. **20** (1984), 237–266.

[67] Huisken, G.: *Asymptotic behavior for singularities of the mean curvature flow*. J. Diff. Geom. **31** (1990), 285–299.

[68] Ilmanen, T.: *Convergence of the Allen–Cahn equation to Brakke's motion by mean curvature*, J. Diff. Geom. **38** (1993), 417–461.

[69] Ilmanen, T.: *Elliptic regularization and partial regularity for motion by mean curvature*, Memoirs Amer. Math. Soc. 108 no. 520 (1994).

[70] Ilmanen, T.: *Singularities of mean curvature flow of surfaces*, preprint.

[71] Jäger, W. – Kaul, H.: *Uniqueness and stability of harmonic maps, and their Jacobi fields*. Manuscr. math. **28** (1979) 269–291.

[72] Jerrard, R. L. – Soner, H. M.: *Scaling limits and regularity results for a class of Ginzburg–Landau systems*, preprint (1995).

[73] Jörgens, K.: *Das Anfangswertproblem im Grossen für eine Klasse nicht-linearer Wellengleichungen*, Math. Z. **77** (1961), 295–308.

[74] Jost, J.: *Ein Existenzbeweis für harmonische Abbildungen, die ein Dirichletproblem lösen, mittels der Methode des Wärmeflusses*. Manusc. Math. **34** (1981), 17–25.

[75] Kapitanski, L.: *Global and unique weak solutions of nonlinear wave equations*, Math. Research Letters **1** (1994), 211–223.

[76] Klainerman, S.: (private communication).

[77] Klainerman, S. – Machedon, M.: *Space-time estimates for null forms and the local existence theorem*, Comm. Pure Appl. Math. **46**:9 (1993), 1221–1268.

[78] Klainerman, S. – Machedon, M.: *Finite energy solutions of the Yang–Mills equations in $\mathbb{R}^{3+1}$*, Ann. Math. (2) **142**:1 (1995), 39–119.

[79] Klainerman, S. – Machedon, M.: *On the regularity properties of a model problem related to wave maps*, Duke Math. J. **87**:3 (1997), 553–589.

[80] Ladyženskaya, O. A. – Solonnikov, V. A. – Ural'ceva, N. N.: *Linear and quasilinear equations of parabolic type*. AMS, Providence R. I. 1968.

[81] Lin, F.-H.: *Une remarque sur l'application $x/|x|$*. C. R. Acad. Sc. Paris **305** (1987), 529–531.

[82] Lions, J.-L.: *Quelques méthodes de résolution des problèmes aux limites non linéaires*, Dunod, Gauthier–Villars, Paris 1969.

[83] Müller, S. – Struwe, M.: *Global existence of wave maps in $1+2$ dimensions for finite energy data*, in preparation.

[84] Müller, S. – Struwe, M.: *Spatially discrete wave maps on $(1+2)$-dimensional space-time*, Topol. Methods Nonlinear Anal., to appear.

[85] Osher, S. – Sethian, J. A.: *Fronts propagating with curvature-dependent speed: Algorithms based on Hamilton–Jacobi formulations*. Journal of Computational Physics **79** (1988), 12–49.

[86] Pecher, H.: *Ein nichtlinearer Interpolationssatz und seine Anwendung auf nichtlineare Wellengleichungen*, Math. Z. **161** (1978), 9–40.

[87] Råde, J.: *On the Yang–Mills heat equation in two and three dimensions*, J. Reine Angew. Math. **431** (1992), 123–163.

[88] Rivière, T.: *Régularité partielle des solutions faibles du problème d'evolution des applications harmoniques en dimension deux*, preprint, 1992.

[89] Rivière, T.: *Everywhere discontinuous Harmonic Maps into Spheres*, to appear, and, *Applications harmoniques de B^3 dans S^2 partout discontinues*, C. R. Acad. Sci. Paris, **314** (1992), 719–723.

[90] Sacks, J. – Uhlenbeck, K.: *The existence of minimal immersions of 2-spheres*. Ann. of Math. **113** (1981), 1–24.

[91] Schiff, L. I.: *Nonlinear meson theory of nuclear forces*, I, Phys. Rev. **84** (1951), 1–9.

[92] Schlatter, A.: *Global existence of the Yang–Mills flow in four dimensions*, J. Reine Angew. Math. **479** (1996), 133–148.

[93] Schlatter, A. – Struwe, M. – Tahvildar-Zadeh, A.: *Global existence of the equivariant Yang–Mills heat flow in four space dimensions*, Amer. J. Math. **120** (1998), 117–128.

[94] Schoen, R.: *Conformal deformation of a Riemannian metric to constant scalar curvature*, J. Differential Geometry bf 20 (1984), 479–495.

[95] Schoen, R. S. – Uhlenbeck, K.: *A regularity theory for harmonic maps.* J. Diff. Geom. **17** (1982), 307–335, **18** (1983) 329.

[96] Schoen, R. S. – Uhlenbeck, K.: *Boundary regularity and the Dirichlet problem for harmonic maps.* J. Diff. Geom. **18** (1983), 253–268.

[97] Schoen, R. M. – Yau, S.-T.: *Conformally flat manifolds, Kleinian groups and scalar curvature*, Invent. Math. **92** (1988), 47–71.

[98] Segal, I. E.: *The global Cauchy problem for a relativistic scalar field with power interaction*, Bull. Soc. Math. France **91** (1963), 129–135.

[99] Shatah, J.: *Weak solutions and development of singularities in the $SU(2)$ σ-model.* Comm. Pure Appl. Math. **41** (1988), 459–469.

[100] Shatah, J. – Struwe, M.: *Regularity results for nonlinear wave equations*, Annals of Math. **138** (1993), 503–518.

[101] Shatah, J. – Struwe, M.: *Well-posedness in the energy space for semilinear wave equations with critical growth*, Inter. Math. Res. Notices **7** (1994), 303–309.

[102] Shatah, J. – Tahvildar-Zadeh, A.: *Regularity of harmonic maps from the Minkowski space into rotationally symmetric manifolds*, Comm. Pure Appl. Math. **45** (1992), 947–971.

[103] Shatah, J. – Tahvildar-Zadeh, A.: *On the Cauchy problem for equivariant wave maps*, Comm. Pure Appl. Math. **47** (1994), 719–754.

[104] Simon, L.: *Singularities of geometric variational problems*, in: *Nonlinear partial differential equations in geometry* (eds. R. Hardt and M. Wolf), IAS/Park City Math. Ser. 2 (1996), 185–223.

[105] Simon, L.: *Proof of the basic regularity theorem for harmonic maps*, in: *Nonlinear partial differential equations in geometry* (eds. R. Hardt and M. Wolf), IAS/Park City Math. Ser. 2 (1996), 225–256.

[106] Soner, H. M.: *Motion of a set by the curvature of its boundary*, J. Differential Equations **101** (1993), 313–371.

[107] Soner, H. M. – Souganidis, P. E.: *Uniqueness and singularities of cylindrically symmetric surfaces moving by mean curvature*, Comm. Partial Differential Equations **18** (1993), 859–894.

[108] Struwe, M.: *On the evolution of harmonic maps of Riemannian surfaces.* Math. Helv. **60** (1985), 558–581.

[109] Struwe, M.: *Globally regular solutions to the u^5 Klein–Gordon equation.* Ann. Scuola Norm. Pisa **15** (1988), 495–513.

[110] Struwe, M.: *On the evolution of harmonic maps in higher dimensions.* J. Diff. Geom. **28** (1988), 485–502.

[111] Struwe, M.: *The Yang–Mills flow in four dimensions*, Calc. Var. **2** (1994), 123–150.

[112] Struwe, M.: *Geometric evolution problems*, in *Nonlinear Partial Differential Equations in Differential Geometry* (eds. Hardt, R. and Wolf, M.), IAS/Park City Math. Ser. **2** (1996), 257–339.

[113] von Wahl, W.: *The continuity or stability method for nonlinear elliptic and parabolic equations and systems*, Rend. Sem. Mat. Fis. Milano **62** (1992), 157–183 (1994).

[114] Wente, H. C.: *Large solutions to the volume constrained Plateau problem*, Arch. Rat. Mech. Anal. **75** (1980), 59–77.

[115] Ye, R.: *Global existence and convergence of Yamabe flow.* J. Geom. **39** (1994), 35–50.

[116] Zhou, Y.: *Global weak solution for the (2+1)-dimensional wave maps with small energy*, preprint (1996).

[117] Zhou, Y.: *Monotonicity and uniqueness of weak solutions of $1+1$ dimensional wave maps*, Math. Z., to appear.

MATHEMATIK, ETH-ZENTRUM, CH-8092 ZÜRICH, SWITZERLAND
E-mail address: struwe@math.ethz.ch

Participants of the Conference

Jürg Fröhlich

Ehud Hrushovski

Dusa McDuff

Jürgen Moser

Thomas Wolff

Clifford Taubes

John Milnor

Richard Hamilton

Gerd Faltings

Ed Witten

Michael Struwe

Donald C. Spencer

Yum-Tong Siu

Mikhael Gromov

Robert Langlands

Henryk Iwaniec

Small Instantons In String Theory

Ed Witten

My goal today is to answer the question: "What happens when an instanton becomes small?" Of course the question needs some explanation. For it to make sense we need to explain the context in which we are trying to decide what happens when an instanton becomes small. And in another sense what I am really trying to explain is the context. I decided I was going to try to give you in my lecture today some kind of introduction to the physics that I think is really exciting. But having seen many past attempts, including my own, to do essentially just that, I decided that I was not going to tackle the problem head on by writing down Planck's constant and telling you what we do with it and so on. Instead I am going to try to explain physics in the context of what some of you might perceive as a real life problem. So much of the talk will really be explaining what the question means on several levels. And probably we will not do justice to answering it; we will spend most of our time with preliminaries.

First of all, for those of you who do not work with instantons, let me say this. You do not need to know very much about instantons for today's lecture. All you need to know is the following: we are in four dimensions; the four-manifold M on which we are working could be flat $\mathbb{R}^4$, because most of our discussion today will have to do with local properties, but it also could be a more general four-manifold. We consider a vector bundle $V \to M$ over M, with structure group a compact Lie group G, and we let A be a connection on V. This connection has a curvature $F(A)$; $F(A)$ is a two-form with values in the Lie algebra. In four dimensions one can decompose a two-form into self-dual and anti-self-dual parts, in this case $F(A) = F^+(A) + F^-(A)$. The famous instanton equation

$$F^+(A) = 0$$

says that half the curvature vanishes. This nonlinear equation for the connection has one very basic property that you really need to know for today's lecture: the equation is conformally invariant, which means that it uses a conformal structure on the four-manifold, not a metric. A special case of conformal invariance is that the equation is invariant under dilations of space. If we consider instantons on flat $\mathbb{R}^4$ with coordinates $\mathbf{x} = (x_1, \ldots, x_4)$, then conformal invariance means, among other things, that the equation is invariant under multiplying all the coordinates by a common scalar, So instantons on flat $\mathbb{R}^4$ come in all sizes. This is indicated in Figure 1, in which the instantons sketched in parts (a), (b), (c) are conformally the same. In the figure, the horizontal direction is meant to be $\mathbb{R}^4$ and the vertical

1991 *Mathematics Subject Classification.* 81T30, 81T60.

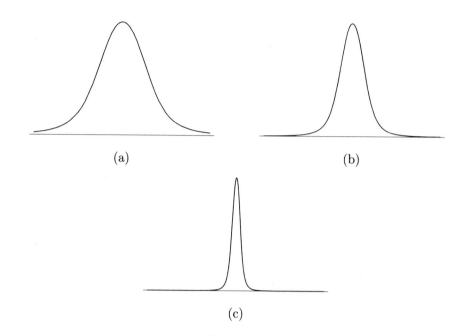

(a)   (b)

(c)

FIGURE 1

direction is a function on $\mathbb{R}^4$ that tells you where the instanton is. An instanton has an energy or action density—mathematicians call it energy but to physicists it is action—defined as $\operatorname{tr} F \wedge F$. The total action integral

$$I = \int_{\mathbb{R}^4} \operatorname{tr} F \wedge F$$

always gives $8\pi^2$, but how the action density is distributed in $\mathbb{R}^4$ depends on what instanton solution you pick. This pretty much summarizes what you should know about instantons as background to today's lecture.

Now, it is clear in physics that the process of scaling down an instanton to make it smaller and smaller cannot go on indefinitely. The classical instanton equations are an approximation to physics but they are not all there is. They are good for slowly varying big things but they would ultimately break down for sufficiently small things. The instanton equations were first studied in the 1970's as an approximation to the theory of the nuclear forces and in that context the instanton equation is an approximation to what you should do in nature. I'll tell you more about better approximations later. The physical question from that point of view is how you should modify this description when the instanton gets so small that the instanton equation is no longer a good approximation to nature.

If you are not interested in physics but only in the mathematical instanton equation, then as this equation is conformally invariant, you can scale the solution all the way down until you reach a sort of δ function singularity. The moduli space $\mathcal{M}$ of instantons has a natural Riemannian structure, and in that structure it is not a complete manifold—in a finite distance you can reach the δ function: the moduli space actually has a singularity. In Figure 2, I have sketched the moduli space in a very schematic way. It has a conical singularity which is associated with the problem we're discussing today of very small instantons.

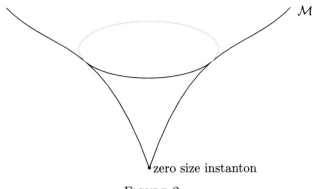

zero size instanton

FIGURE 2

What are we supposed to say about the singularity? Of course that depends on what we are interested in doing with instantons. First of all, what is the traditional answer? Traditionally in instanton physics since the middle and late seventies, we were studying instantons because we were trying to study a certain quantum field theory, and because some questions in that theory gave us a measure on the smooth part of the instanton moduli space which we wanted to integrate. The traditional physics answer was to show that the integral over the smooth part of the moduli space converges. Asymptotic freedom usually enters in proving that. Once this is established, one simply ignores the singularities. And that's the right answer for those questions.

On the other hand some of you might think of instantons as a tool for studying the geometry of four-manifolds in the context of Donaldson theory. In that case you are basically not integrating a measure on $\mathcal{M}$; you are doing some kind of intersection theory on $\mathcal{M}$. In doing intersection theory, we cannot just ignore the singularities. We either work very hard to keep away from the singularities or else we somehow take them into account. In Donaldson theory, there is quite an elaborate story of how to take into account these and other singularities in the instanton moduli space; this is probably the main source of technical difficulty in the theory.

Fast variables and slow ones

So far I have given you the traditional physics answer about how to deal with small instantons, and also the traditional math answer. But there is a different physics answer, which is what I want to focus more on today, and that has to do with studying instantons in real physics—not in the idealized model where the equations are scale invariant, but in a description which is a better approximation to nature. And that will be string theory, where the scale invariance of the classical instanton equation, or the conformal invariance as I called it before, will be a good description for big instantons but will break down for sufficiently small instantons. In that case as we've learned in the last year, string theory gives a completely different kind of answer to the question: "What happens when an instanton is small?"

I first will aim to orient you to the kind of answer we will head for. We will do that by thinking about a much more elementary problem. We are going to go back to classical mechanics, and we'll think about a particle moving in N-dimensional

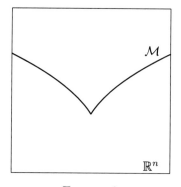

$$\mathcal{M}$$

$$\mathbb{R}^n$$

FIGURE 3

Euclidean space, $\mathbb{R}^N$—we could take a more general Riemannian manifold, but flat Euclidean space will be good enough. On this Euclidean space we put a potential energy, $V(x_1, \ldots, x_N)$. We take this potential energy to be smooth and non-negative, $V \geq 0$, and we'll suppose that $V = 0$ defines a sub-manifold $\mathcal{M} \subset \mathbb{R}^N$, which I'll call $\mathcal{M}$ because it will enter our mechanics problem in a way somewhat similar to the role of instanton moduli space, to which I gave the same name (see Figure 3). We will assume that at a generic point of $\mathcal{M}$, the equation $V = 0$ defines this manifold as non-degenerately as it can, given the hypotheses about V. To be precise, we assume that at a generic point in $\mathcal{M}$, V vanishes precisely to second order in the normal directions and that the matrix of second derivatives in the normal direction (sometimes called the Hessian) is non-degenerate. To be more precise, the matrix of second derivatives will be non-degenerate at smooth points of $\mathcal{M}$, but we assume that there may be singularities, at which the matrix of second derivatives is degenerate. In this situation now we want to look at Newton's laws of classical motion,

$$\frac{d^2 x_i}{dt^2} = -\frac{\partial V}{\partial x^i}$$

and also at the quantum version.

But we are going to look only at motions of very low energy. Such a restriction often arises for practical reasons. For example, if you are an experimentalist in the laboratory and you simply don't have the equipment to create initial conditions where you could see the motions of high energy, then you look at the motions of low energy because you have no choice. Since the energy is low, we are going to be very close to $\mathcal{M}$. We are essentially trapped on $\mathcal{M}$, and this problem can be systematically treated.[1] What you will see will be approximately geodesic motion on $\mathcal{M}$. In general, $\mathcal{M}$ can be an interesting Riemannian manifold which is embedded in Euclidean space and gets some induced metric, and the leading approximation is the geodesic motion on $\mathcal{M}$ in its induced metric. Then, if you want, you can calculate a corrections to that approximation.

To describe this more systematically, if we're at a smooth point in $\mathcal{M}$, then locally we have coordinates, which I'll call u^α, which are tangent to $\mathcal{M}$, and we have also the normal coordinates, denoted y^γ. The motion along $\mathcal{M}$ will be slow, because the energy is small, but the motion in the normal direction will actually

---

[1] For the rather analogous problem of motion in a rapidly varying field see L. D. Landau and E. M. Lifschitz, *Mechanics* (Pergamon Press, 1960), pp. 93–5.

be a very fast vibration, as in the case treated in the reference of footnote 1. The slow variables describe the motion along $\mathcal{M}$, and in the leading approximation you just see the geodesic equation

$$(1) \qquad \frac{D^2 u^\alpha}{Dt^2} = 0.$$

But if you average more precisely over the fast variables to track the equations for the slow variables, you'll find some higher order corrections that you can work out systematically. Generically, these corrections may induce an effective potential energy on $\mathcal{M}$.

The closest analog of our instanton problem is a situation in which (perhaps because of supersymmetry and quantum mechanics) such a potential energy is not induced. Even if no potential is induced, averaging over the fast variables will still correct the effective equations for the slow variables by correcting the effective Riemannian metric on $\mathcal{M}$ and by inducing terms of higher than second order in time derivatives. The effective equations will thus look something like

$$\frac{D^2 u^\alpha}{Dt^2} = X^\alpha_\beta(u) \frac{D^4 u^\beta}{Dt^4} + \cdots$$

with some X^α_β. (The "..." are terms of still higher order in time derivatives. Because of the microscopic time reversal symmetry of Newton's laws, I have omitted in this equation terms of odd order in D/Dt.)

I now will make a digression to explain something that is important although somewhat removed from our main topic. Classically, the corrections to the geodesic equation on $\mathcal{M}$ depend on the energy stored in the fast variables, which is approximately conserved. This is actually a point where the quantum description of a system such as this one is significantly different from the classical description. Quantum mechanically, near smooth points of $\mathcal{M}$, the nondegeneracy of the Hessian means that there is a unique quantum state of lowest energy for the motion in the normal directions. Let E_0 be the ground state energy of the Hamiltonian that governs motion in the normal directions, and let E_1 be the energy of the first excited state of that Hamiltonian. Both E_0 and E_1 should be regarded as functions on $\mathcal{M}$. E_0 can be interpreted as an effective potential for the motion on $\mathcal{M}$. To get a good quantum mechanical analog of our instanton problem, we really should consider a system for which the function E_0 is identically zero (or at least constant) so that a low energy particle can move freely on $\mathcal{M}$. This rather exceptional behavior occurs most naturally in a supersymmetric extension of the system under consideration here, but I will not pause to clarify that point. The function $E_1 - E_0$ is strictly positive away from singularities of $\mathcal{M}$. As long as one considers motions of energy much less than E_1, the "fast" motions in the normal directions to $\mathcal{M}$ are locked in their ground state. Under these conditions, the quantum system reduces to the quantum version of geodesic motion on $\mathcal{M}$ (or its supersymmetric extension). The effective Hamiltonian is the Laplacian on $\mathcal{M}$.

The difference that I have tried to convey between the classical and quantum mechanical treatment of the fast variables actually has something to do with how quantum mechanics was discovered historically. Planck's 1900 paper on black body radiation and Einstein's 1905 paper on the specific heat of solids both dealt with situations in which the motion of the fast variables was suppressed experimentally, in ways that were not expected classically.

Let us go back to the equation that we had before our quantum mechanical digression. One important point about this equation is that the geodesic equation is second order in time, but if one wishes to improve upon the geodesic equation to get a more precise description of the slow motion, one must add to the geodesic equation terms of higher than second order. This is dangerous, because the corrected equation with such higher order terms added has more solutions than the original geodesic equation; a larger number of initial conditions must be specified in order to determine a solution. The geodesic equation with the corrections added should therefore not be taken too literally; the corrections should only be evaluated perturbatively (and to incorporate the effects of a more and more precise averaging over the fast variables, more and more correction terms need to be taken into account). To get an exact treatment, one would have to go back to the original equations for motion on $\mathbb{R}^N$; these equations of course describe additional degrees of freedom—the fast variables—beyond what is described by the geodesic equation on $\mathcal{M}$.

I have gone into so much detail about this because the situation I have described is almost universal in physics. There almost always is something you do not know about. You do not know about it because it is vibrating too fast or because the energies involved are too big. So in practice you are always averaging over some fast variables, and the equations you study are the equations that you get when you average out the fast variables. If you study such a system more precisely, what generally happens is that any equations you may have written involving only the slow variables need to be corrected, in a more accurate description, by adding additional terms that come from a more precise process of averaging over the fast variables. The corrections always have the dangerous character of too many derivatives or some other, related, bad properties. When you find that treating a system more precisely requires introducing in the equations new terms that have bad properties, this usually means that you have been, perhaps unknowingly, integrating out some fast variables.

So for instance, hundreds of years ago Euler and others studied motion of a rigid body, and there were all kinds of fast variables that they did not know about. They did not know about electrons, protons, and neutrons, not to mention quarks and gluons and so on. Much of what was discovered in the last century were fast variables that people who traditionally studied rigid motion did not know about. Nevertheless, Euler and his contemporaries had a great deal of success in studying rigid body motion, writing equations which are valid once one averages over the fast variables.

To give another example that is perhaps closer to our topic today, a century ago beta decay of atomic nuclei was discovered, and in the 1930's Fermi wrote down the first really successful quantum theory of beta decay. To do so, Fermi added to the equations (or better, the Lagrangian) governing electrons, neutrinos, and other particles additional terms that worked fine if treated perturbatively, but which if taken too seriously were actually badly behaved. The bad behavior was somewhat analogous to the bad behavior of the higher derivative corrections to the geodesic equation that were mentioned above. This bad behavior was ultimately understood as a clue that Fermi's equations result from averaging over the unobserved values of some previously unknown fast variables. What is now the standard model of electroweak interactions was created when Weinberg, Glashow, and Salam guessed what the fast variables were. They were in fact the W and Z bosons that were discovered experimentally at CERN in the early 1980's, and a Higgs boson (or possibly several particles whose effects can be imitated at lower energies by a single Higgs boson) whose existence has not yet been confirmed experimentally.

Learning about the fast variables

Unless one knows everything, a state of affairs that does not ordinarily prevail, there are always some fast variables one does not know about that one is implicitly averaging over. In this sense, the situation I am describing is universal in physics.

Learning about the fast variables which we do not know about and which our predecessors have averaged over is one of the main goals of physics, one of the main goals of experiments. In that branch of physics which is my main interest, where the focus is to learn the fundamental laws of nature, at the risk of perhaps some slight oversimplification, we can say that the universal problem is to learn what are the fast variables over which physicists have integrated or averaged so far.

There are some obvious strategies for doing so. You can ask Congress to appropriate the necessary funds so that you can increase the energy, and when you increase the energy the fast variables will be more obvious. There might be some even faster ones that you still won't see, but if you increase the energy at least some of the fast variables might become more important.

Alternatively, I explained that even at low energies, there are corrections to the geodesic motion, and you could also (although this might again require asking Congress for additional funds) increase the precision of your measurements while keeping the energy fixed, and see the effects of fast variables even at low energies. From sufficiently precise measurements even at fixed energies you could infer something about the fast variables. So for instance, in an example I gave above, for the first ninety years after the discovery of beta decay—until the W boson was discovered at CERN in the 1980's—that's roughly what was done. Physicists studied the effects of the W boson by accurate measurements at low energies, without having the tools to observe W bosons directly.

But the formulation of the problem that I gave earlier might suggest another approach to a mathematician. We could go to a singularity of $\mathcal{M}$ where the Zariski tangent space increases because the matrix of second derivatives of the potential in the normal directions is no longer non-degenerate.

At such a point, some of the fast variables become slow and obvious. In the quantum mechanical version of the story, the function $E_1 - E_0$ vanishes at such a point, and at least some of what formerly were the fast variables are no longer frozen out. If you go to a singularity at least some of the things you don't know about become clear. That's a hard strategy to follow in real experiments but it is one of the main strategies used in thought experiments in string theory in the last year. In string theory, which physicists have been studying for the last twenty-five years or so, there is a fantastically rich structure which I think I can tell you probably really is physics. It is probably the successor to ordinary geometry for physics and perhaps also for math, but large parts of the puzzle are not yet clear at all even at the most basic level. For one thing, in a certain sense we have only been studying the slow variables for the last twenty-five years. (This assertion is more obvious now than it was a year or two ago, before the new discoveries about string duality, D-branes, and the like.)

We are, as theorists, in a situation a little bit similar to the situation that experimentalists are usually in. In string theory our predecessors of the late 1960's and 1970's stumbled upon a theory that has been intensively studied every since, but which is very far from being fully understood. We know some pieces of the puzzle, and we do not know others. To try to probe some of the bits that we do not understand, we perform thought experiments that have some analogies with real experiments.

There are many differences between thought experiments and real experiments, of course. For today, the main difference is that the third strategy for learning about the fast variables—the strategy that is not really practical for experimentalists—has turned out to be quite practical and useful for thought experiments in string theory. I'll now summarize a point of view about string theory which has been confirmed by a lot of recent and less recent discoveries.

Singularities and surprises

In string theory there are no singularities, only surprises. If something appears to be a singularity in some approximation then if you look at it more closely you simply learn about a new piece of the structure that you may not have suspected at all.

By contrast, singularities often do occur in ordinary geometry. The singularity I mentioned at the beginning was the singularity when an instanton tries to turn into a δ function. Ordinary geometry is sometimes a good approximation to the new geometry which is its successor in string theory, but it sometimes isn't, and if you try to go to a singularity or more exactly if you try to go to what a traditional geometer would call a singularity, you always will find that something different happens in string theory. Some of the fast variables that you may not have known about become slow and a new piece of the puzzle pops out. And you find the definite physical mechanism—or mathematical mechanism—that depends on what used to be fast variables which become slow variables at the singularity. Lots of things along these lines have become clear lately.

Before trying to focus on the small instanton example. I want to briefly mention a few other examples. One of them will be the A–D–E singularities of complex surfaces. For some of you this may be a familiar subject. Here you are dealing with a complex surface which, if it is smooth, looks locally like a piece of $\mathbb{C}^2$. In a family of such surfaces you may very naturally get certain kinds of singularities which have a lot of distinguished properties and have fascinated mathematicians for much of the last century if not more. They are the isolated singularities in two dimensions that do not contribute to the canonical divisor. They have the property that when you resolve them you find configurations of rational curves arranged according to Dynkin diagrams of type A–D–E.

Mathematicians call these the A–D–E singularities and thought of them as somehow being connected with A–D–E groups. But at least to an outsider it looks like what was really seen mathematically was a lot more abstract then the group. Dynkin diagrams are all very well and good but they are a method of studying groups, not the other way around, at least for some of us. It is not clear to me that in classical geometry you see the group in a fashion that is anywhere near as direct as string theorists came to understand it in the last year, when we learned that Type IIA superstring theory, on a complex surface that develops an A–D–E

singularity, gets an A–D–E gauge group. If the singularity we were discussing were the A–D–E singularity and the question was: "What does Type IIA string theory do at an A–D–E singularity?" the answer is: "It gets an A–D–E gauge group." That is an example of what I mean by saying that in a situation where classical geometry produces a singularity, string theory produces a physical or mathematical mechanism rather than a singularity.[2]

For a second example, instead of singularities of compact surfaces, consider singularities of complex threefolds. Here the classification of the interesting singularities to study is much more complicated and there is a much richer structure of answers. The full story is not known, but we certainly do know the answer for many types of singularities. For instance, for an ordinary double point of a complex threefold, $x^2 + y^2 + z^2 + w^2 = 0$, the physical mechanism turns out to be the appearance of a "massless hypermultiplet."[3]

We want an answer in a similar spirit for the problem of a small instanton. When an instanton becomes small some new degrees of freedom should materialize. It could be that some manifestations of these new degrees of freedom are seen in classical geometry, as in the case of the A–D–E singularities of surfaces, where the appearance of a special configuration of rational curves in the resolution is related to the fact that the string theory gets a gauge group in the case of the A–D–E singularity. So you might ask yourself: "What bit of classical geometry might materialize and turn into physics when one considers a small instanton?" But I'm not going to pose this as an examination question today. I'll just tell you the answer:[4] the fast variables that become slow are the variables that appear in the so-called ADHM construction of instantons, [5] which becomes physical and which you would have to discover in doing the string theory if it weren't already known. Explaining this statement as well as I can is the goal of the rest of this talk. After describing some background, we will come back to the small instanton problem at the end.

Two parameter deformation of geometry

Well, what is string theory? String theory is roughly a two parameter deformation of the world, or at least a two parameter deformation of classical physics and geometry.

In the lower left-hand corner of Figure 4 I have put classical partial differential equations (P.D.E.'s), as they appear in classical field theory and in traditional Donaldson theory. We can deform the classical P.D.E. in two directions to better describe physics. We can turn on Planck's constant $\hbar$ and that gives us the quantum field theory. Or we can turn on α', as it is called, a parameter with dimensions of length squared which appears in going to string theory. (Numerically α' is very

---

[2]Unfortunately, there is not a reference explaining for mathematicians the string theory interpretation of the A–D–E singularity. The appearance of the gauge group was noted in E. Witten, "String theory dynamics in various dimensions," Nucl. Phys. **B443** (1995) 85, and further elaborated in P. Aspinwall, "Enhanced gauge symmetries and K3 surfaces," Phys. Lett. B357 (1995) 329. A mathematical reference giving background to some of these matters is D. Morrison, "The geometry underlying mirror symmetry," alg-geom/9608006.

[3]A. Strominger, "Massless black holes and conifolds in string theory," Nucl. Phys. **B451** (1995) 96.

[4]See E. Witten, "Small instantons in string theory," Nucl.Phys. **B460** (1996) 541-559.

[5]M. F. Atiyah, V. G. Drinfeld, N. Hitchin, and Yu. I. Manin, "Construction of instantons," Phys. Lett. **A65** (1978), 185.

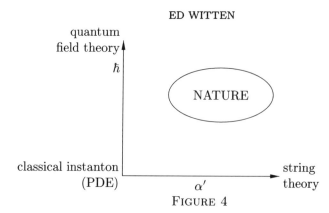

FIGURE 4

small by usual standards, about $(10^{-32}\,\text{cm})^2$ according to the traditional way of identifying string theory parameters with experimental ones.)

In a sense mathematicians know something about $\hbar$ but nothing at all about α'. First of all quantum field theory is an old subject by now—it was initiated in the 1920's—and mathematicians have grappled with it to some extent. For instance, in constructive field theory there are extensive existence theorems about quantum field theories in many simple and not so simple cases.[6] Moreover, geometers have been increasingly interested in results coming from quantum field theory in the last ten or twenty years.

But this is only part of the experience that mathematicians have with quantum field theory. One may fairly state that the one-dimensional case of quantum field theory is much of twentieth century mathematics. Such subjects as representation theory, harmonic analysis, and elliptic operators, lots of things that have been important in twentieth century mathematics, deal with quantum field theory in one space-time dimension. But this one-dimensionality is not generally perceived in standard mathematical treatments. What is one-dimensional in much of conventional mathematics is what physicists call the world-line, which is usually suppressed completely in the mathematical discussions; the target space, on which one formulates an elliptic operator upon quantization, can have any dimension in one-dimensional quantum field theory. So the vertical or quantum mechanical deformation in my figure has been studied extensively by mathematicians in the twentieth century in the one-dimensional case, though generally without thinking about it in that way.

And in higher dimensions mathematicians have at least grappled with quantum field theory to some extent, though you have not come to grips with it as you probably will attempt to in the next century.

The α' deformation, however, is a different kettle of fish; mathematicians have not even begun to grapple with it at all, and it is generally not even appreciated that doing so might be one of the eventual goals of mathematical research. But as I have said before, we have good reasons at least to suspect that nature is somewhere out in the first quadrant. If that turns out to be so, then sooner or later, maybe far in the next century, the full two parameter deformation of classical geometry is bound to turn out to be important for pure mathematics.

Today I really don't have time to explain the deformation of classical geometry that I have depicted vertically in the figure—the quantum field theory deformation.

---

[6]For a comparatively recent treatment see V. Rivasseau, *From Perturbative To Constructive Renormalization* (Princeton University Press, 1991).

I will have to pretend that that is familiar. We will concentrate on the horizontal or string theory deformation. Here I come to the one theorem that I will prove today: it is impossible to give the lecture I want to give. In fact the basic prerequisite for understanding the horizontal deformation is to first understand the vertical one; and since it takes an extensive course in quantum field theory to properly explain what the vertical deformation means, we really cannot properly proceed with the lecture. We will attempt to ignore that and to describe the horizontal deformation anyway.

String theory deformation of an instanton

So, accepting the inevitable limitations, I'm going to start by explaining what happens to an instanton when you move horizontally.

When you move horizontally into the string theory world, everything you know gets corrected. We do not really understand it properly because we don't know the fast variables. As I told you, if one does not know about the fast variables, one can still integrate them out, average over them and get equations for slow variables. But the price one pays is not completely understanding what one is doing. The results that arise by averaging over unknown fast variables, though not fully satisfactory, can be very concrete. For instance, important equations in physics and geometry are the Einstein equations, $R_{IJ} = 0$, where R is the Ricci tensor. In string theory, upon averaging over the fast variables you get corrections to the Einstein equations. You can systematically calculate the leading corrections and they look something like this:

$$R_{IJ} = \alpha' \left(R_{IKLM} R_J^{KLM} + \Box R_{IJ} \right) + \cdots .$$

Here α' is the new constant of nature, mentioned earlier, that appears in string theory. In this equation, since the Einstein equations are of second order and the terms on the right hand side are of higher than second order, we see the traditional dangerous structure with too many derivatives that comes from averaging over fast variables instead of understanding them. The string theory correction to the Einstein equations that I have schematically written down is not experimentally testable in practice but it is experimentally testable in principle. In other words, in principle, in string theory one can compute definite corrections to the Einstein equations which make corrections in the solar system, but these corrections unfortunately are unmeasurably small. String theory reduces only at long distances to the Einstein equations. It really is something different in general.

The instanton becoming small will be a situation where the corrections are big. The solar system is a situation where the corrections are small.

So far I have overlooked some crucial fine structure. Among the fast variables in string theory there are those that we've known about for decades, which have to do with vibrations of a string, and those that we've learned about only in the last year by going to singularities. The corrections to the Einstein equations that I wrote schematically above, for instance, are part of the "old" story, with which one really must be familiar to be able to appreciate the newer results.

Those of the fast variables that have been known traditionally can be described in a rather systematic framework. That framework is called *conformal field theory*. Conformal field theory is something mathematicians have grappled with a bit, thinking of it as a special case of the vertical deformation (or sometimes just as a specialized branch of representation theory). What is much deeper and what I think mathematicians have not encountered at all is in the use of conformal field theory in the horizontal deformation. That's a lot harder to explain, but it is what I want to try to convey next.

To do this, we just have to change our way thinking about everything. Given any data, for example a metric g on a manifold M, we are going to use it to construct an auxiliary object which is a two-dimensional field theory. With metrics one can do all kind of things. What we're going to do with metrics is to write down a Lagrangian functional. So, given the metric g on M we introduce an auxiliary two dimensional surface Σ, a two real dimensional surface. You could call it a Riemann surface if the appropriate structure is present. Then we consider a map

$$X : \Sigma \longrightarrow M$$

and we let I be the area of $X(\Sigma)$ in M. And once you have got such an I you formally take the step which I would be focusing on if I were really describing the vertical or quantum deformation. Once you have defined such an I now you try to integrate over the space of all maps weighted by e^{-I}

$$Z = \int \mathcal{D}X \, e^{-I}$$

Making sense of this gives what's called a two-dimensional quantum field theory. This gives the proof of the one theorem I stated earlier. To understand the horizontal (stringy) deformation one must first understand the vertical (quantum) deformation, because when I try to explain the horizontal deformation practically the first thing that happens is that we run into an example of the vertical deformation. I can't really explain it today so we have to treat it as a black box.

In sum, given a metric g on a manifold M, instead of doing any of the traditional things that geometers do with metrics, what you now do is to use the metric to construct an auxiliary two dimensional problem which you then have to study quantum mechanically. Then you can get a dictionary where traditional ideas in geometry get modernized at least for physics. The dictionary may not be complete since some traditional notions do not carry over nicely to string theory.

Here are some important examples of the dictionary. The analog of what in geometry is a metric (and collection of other fields in space-time) is a two-dimensional field theory. The analog of a metric that obeys the Einstein equations (or a collection of a metric and other fields that obey the appropriate classical field equations) is a conformal field theory.

Here are some additional important examples. A bit of old geometry is the notion of a Kähler metric and the corresponding notion in string theory is what's called $N = 2$ supersymmetry. And hyper-Kähler geometry turns into $N = 4$ supersymmetry.

Why did I mention hyper-Kähler geometry? The reason is this: hyper-Kähler geometry is to metrics as self-duality is to connections. So knowing about how hyper-Kähler geometry enters conformal field theory is actually important background for thinking about the instanton problem.

Incorporating the connection in conformal field theory

Next I would like to include in the discussion the idea of a connection on a fiber bundle so that we can talk about instantons. I've told you that everything is going to turn into data in constructing two-dimensional Lagrangians and that will be true for the connection also. It takes some explanation to tell what kind of Lagrangian you write using a connection. Roughly speaking you've got a vector bundle V with connection A over a Riemannian manifold M. You introduce as before an auxiliary Riemann surface Σ, and consider

$$X : \Sigma \longrightarrow M.$$

Then you pull back the bundle $X^*(V)$ and connection $X^*(A)$ and write a Lagrangian

$$L = \text{ Area } + \int \lambda(d + X^*(A))\lambda,$$

where λ is a (fermionic) section of $X^*(V) \to \Sigma$. In this way, one uses the connection A as part of the data in defining a two-dimensional field theory. The details are not so important for today. The main thing I want to say is that to answer in string theory the question: "What happens when an instanton is small?" the first step is to consider the two-dimentional Lagrangian that depends on the instanton connection in that way that I have roughly indicated.

Including a connection A on $V \to M$ in string theory means writing down and studying a two-dimensional Lagrangian that depends on A. Once we've put the connection into the two-dimensional Lagrangian, the Yang–Mills equations become (approximations to) the conditions for conformal invariance of the auxiliary two-dimensional model.

We should now ask which two-dimensional conformal field theories correspond to instantons. The answer turns out to be that self-duality corresponds to a property of the two-dimensional conformal field theory that is known as $(0,4)$ supersymmetry. Having gotten this far, we at least know what it means to have an instanton in string theory, and we can then ask what happens when an instanton becomes small. So finally, I have described the setting in which we want to answer the question, "What happens when an instanton becomes small?"

The fate of an instanton

When an instanton is big, we can describe it by the traditional instanton partial differential equation or P.D.E. Deforming it horizontally means that we're supposed to do super-conformal field theory instead of classical P.D.E.'s. If you are not familiar with the "super" part but have heard about ordinary conformal field theory, then just call it conformal field theory. We can try to do conformal field theory and learn what happens when an instanton becomes small. That was done, though certainly not rigorously, by Callan, Harvey, and Strominger.[7] The difference betwen the conformal field theory answer and what one gets from the classical P.D.E. is important only when the instanton is small. After all, physicists studied instantons in the seventies without worrying about string theory, so it must be that when the instanton is big enough you don't need string theory.

In an instanton field, as we explained before, the integral of the action density is $8\pi^2$. The big instanton is spread out over a large region. Traditionally with instantons we don't worry about the metric. But string theory is a unified theory of gauge fields and geometry, so the instanton equation in string theory involves the metric as well as the connection. So you have to solve for the metric (Figure 5). In

---

[7]For example, see C. G. Callan, J. A. Harvey, and A. Strominger, "World-brane actions for string solitons," Nucl. Phys. **B367** (1991) 60.

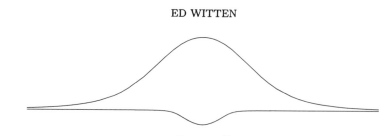

FIGURE 5

a way the main novelty in string theory, or one of the main novelties, is that even if you just want to study instantons, you are forced to worry about geometry as well as gauge theory. You solve for the metric and it is not quite flat. In conventional instanton theory, one simply solves for the instanton connection, but in conformal field theory you solve for the connection and metric together (along with additional data), and for the big instanton you find that the metric has a slight dimple. It's an almost flat metric with a slight dimple.

Now let the instanton become smaller. The action or energy still integrates to $8\pi^2$, but it is localized in a smaller region in space. The metric has a bigger dimple. Figure 6 shows roughly what the dimple looks like.

Now as the instanton shrinks, classically what happens is that the energy becomes a δ function times $8\pi^2$. But a δ function is a singularity, and string theory does not have singularities. So something else will have to happen in string theory. What happens if you try to shrink the instanton is that this dimple develops and becomes longer and longer and as the classical story gets a δ function, this dimple becomes a semi-infinite tube (Figure 7). Instead of a delta function singularity in the action density, as one might expect from the behavior of the classical P.D.E., what happens after we make the horizontal deformation is that space-time develops a second end.

The second end of space-time that appears in this way looks asymptotically like $\mathbf{S}^3 \times \mathbb{R}$. We are doing conformal field theory here so $\mathbf{S}^3 \times \mathbb{R}$ must be understood in conformal field theory and not in classical geometry. You might ask, "What is the conformal field theory of $\mathbf{S}^3 \times \mathbb{R}$?" The main point of the answer is describe the conformal field theory of $\mathbf{S}^3$. This is called the SU(2) WZW model; it is related to the representation theory of the Kac–Moody algebra of SU(2) (the central extension of the loop group of SU(2)). I mention this because the WZW model is something that some of you have met, conceiving of it as a special case of the

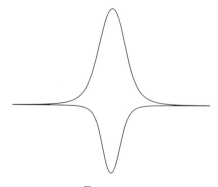

FIGURE 6

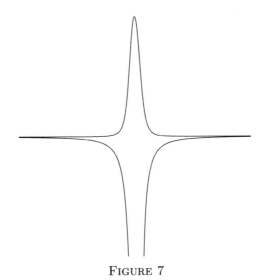

FIGURE 7

vertical deformation or merely as an aspect of representation theory. This may help make more vivid the assertion that the bits of conformal field theory that you may have met in another context are parts of the same subject that is used to generalize geometry to string theory.

But if you are not familiar with the WZW model, never mind. Then the statement is simply that in string theory, where there would have been a singularity is in fact a second end of space. And if you ask what the connection looks like the answer is, it disappears, it's flat. The δ function, so to speak, moves down the end of the tube and disappears. In the classical P.D.E., the limiting connection, for an instanton that shrinks to zero, is flat on the complement of a point where there is a singularity. In string theory the point is deleted and, roughly speaking, is projected to infinity, down the tube.

So this is what a small instanton does on the horizontal axis of Figure 4. We recall that the horizontal axis is what you get upon incorporating string theory but not quantum mechanics. What has so far been described is not, however, the correct answer for nature because nature is not on the horizontal axis. We are somewhere in the first quadrant since quantum mechanics prevails, along with, presumably, string theory.

Quantum mechanics and the ADHM construction of instantons

How far off the horizontal axis are we in nature? That is not a question one can answer in a moment or two because the answer is not simple. For some purposes quantum mechanics is all-important and for other purposes its effects are unmeasurably small. But in our small instanton problem one can make a clean mathematical statement. The statement is that unless the vertical deformation is completely absent (that is, unless one is precisely on the horizontal axis), the vertical deformation becomes important when an instanton shrinks to zero size. In fact, in the course of learning what small instantons do on the horizontal axis, physicists also learned that the horizontal axis can not be a good approximation when the instanton is small. When an instanton is small, it is impossible to turn off quantum mechanics, even as a mathematical exercise. That is because in string theory the smallness or largeness of quantum effects depends on the value of a certain field ϕ (usually called the "dilaton"), roughly as if the effective value of Planck's constant is e^{ϕ}. When ϕ is near $-\infty$, quantum effects are very small, and conformal field theory gives a very good approximation.

In solving for the instanton in conformal field theory, one solves for the metric, the connection, and (among other things) also ϕ. If ϕ is everywhere large and negative, the conformal field theory description is good. The value of ϕ near spatial infinity can be chosen as a boundary condition; if it is, say, -10^{100}, then ϕ is everywhere large and negative as long as the instanton is not too small.

Something else happens (as shown by Callan, Harvey, and Strominger in work I cited earlier) when the instanton becomes small and space-time develops a second end. What happens, in fact, is that ϕ grows linearly as one goes down the tube, and in the limit that space-time has a second end, the value of ϕ at that second end is actually $+\infty$. That is why the vertical deformation is inevitably important in the limit when an instanton becomes small.

What happens when quantum effects turn on, that is when the vertical deformation can no longer be neglected? String theory is probably the richest subject that physicists have ever encountered. But it is also a subject where we only understand pieces, and one of the most fundamental limitations is that traditionally we have only understood what happens near $\phi = -\infty$.

The small instanton problem is an interesting one because the traditional approximation breaks down, but it only breaks down in a controlled way. ϕ is almost everywhere near $-\infty$ but there is a limited region where that fails. Because the traditional situation breaks down only in this limited way, one can actually learn what happens. It is roughly comparable to learning that A–D–E singularities give A–D–E gauge groups. That is another example in which the traditional approximation breaks down only in a very limited region, in fact just near the singularity.

What happens at the A–D–E singularity, or near the small instanton limit, is that some of the variables that one did not understand before, some of the fast variables over which one was always averaging, become slow and obvious. To make sense out of how physics behaves where classical geometry breaks down, one needs to know what fast variables are becoming slow, and then can try to write the appropriate equations incorporating these additional variables. You see in my elementary example of motion in $\mathbb{R}^N$, if you are stuck at a regular point of $\mathcal{M}$ without the energy or precision to see the fast variables, the fast variables are mysterious. But if one can discover what the fast variables are, the appropriate equations are not so mysterious; they are just Newton's laws again with more variables. In the elementary example the mystery is what are the fast variables. If you can guess what they are then you are well on the way to finding the right equations. As I mentioned in the introductory part of this lecture, the new variables in the small instanton case are the variables used in the ADHM construction of instantons.

In the ADHM construction, one aims to construct the instantons and not only the moduli space. For instance, one wants to construct the moduli space $\mathcal{M}_k$ of k-instantons on $\mathbb{R}^4$. Its dimension is $8k$ for SU(2). In the ADHM construction, one describes $\mathcal{M}_k$ as the so-called hyper-Kähler quotient of $\mathbb{R}^{8k+4k^2}$, by a certain linear action of the group $U(k)$. That this auxiliary group $U(k)$ should be introduced in

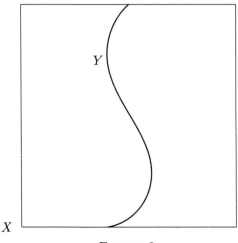

FIGURE 8

the most efficient construction of k-instantons was one of the key aspects of the ADHM construction since its discovery in the 1970's.

To physicists, until the new string theory developments, the ADHM construction was an interesting bit of mathematics, helpful for solving the differential equations but not really integrated into the physics especially at the quantum level. It's now integrated into the physics, though it will perhaps be a bit hard to explain that. Figure 8 is my attempt to do so.

In this figure I'm tossing quite a few new things into the mix. For one thing we're not really in 4 dimensions, we are in a higher dimension (often dimension 10). So X is space-time, of dimension bigger than four. A collection of small instantons are being localized on a codimension-four submanifold Y of X. (A small instanton is localized near a point in $\mathbb{R}^4$, or more generally near a codimension-four submanifold in X.) What in the traditional ADHM construction were simply variables now become fields on Y. The $U(k)$ auxiliary group of ADHM becomes a gauge group along Y. In the same sense that the A–D–E gauge group becomes obvious near an A–D–E singularity, the ADHM construction of instantons becomes obvious near a small instanton singularity.

That is more or less what I have to offer today. I have tried to use the question "What happens when an instanton becomes small?" to expose you to the two fundamental deformations of traditional geometry, of traditional mathematics, that physicists have been grappling with and that you mathematicians may well increasingly grapple with in the time before the 300th aniversary conference of Princeton University. Thank you.

INSTITUTE FOR ADVANCED STUDY, SCHOOL OF NATURAL SCIENCE, PRINCETON, NJ 08540
E-mail address: witten@ias.edu

Recent Work Connected with the Kakeya Problem

Thomas Wolff

A Kakeya set in $\mathbb{R}^n$ is a compact set $E \subset \mathbb{R}^n$ containing a unit line segment in every direction, i.e.

(1) $$\forall e \in S^{n-1} \; \exists x \in \mathbb{R}^n : x + te \in E \; \forall t \in [-\tfrac{1}{2}, \tfrac{1}{2}]$$

where S^{n-1} is the unit sphere in $\mathbb{R}^n$. This paper will be mainly concerned with the following issue, which is still poorly understood: what metric restrictions does the property (1) put on the set E?

The original Kakeya problem was essentially whether a Kakeya set as defined above must have positive measure, and as is well-known, a counterexample was given by Besicovitch in 1920. A current form of the problem is as follows:

Open question 1: Must a Kakeya set in $\mathbb{R}^n$ have Hausdorff dimension n?

When $n = 2$, the answer is yes; this was proved by Davies [19] in 1971. Recent work on the higher dimensional question began with [7]. If $\dim E$ denotes the Hausdorff dimension then the bound $\dim E \geq \frac{n+1}{2}$ can be proved in several ways and may have been known prior to [7], although the author has not been able to find a reference. The recent work [7], [60] has led to the small improvement $\dim E \geq \frac{n+2}{2}$. We will discuss this in section 2 below. (See also the remark at the end of Section 2.)

Question 1 appears quite elementary, but is known to be connected to a number of basic open questions in harmonic analysis regarding estimation of oscillatory integrals. This is a consequence of C. Fefferman's solution of the disc multiplier problem [23] and work of Cordoba (e.g. [18]) and Bourgain (e.g. [7], [9], [10]). We will say something about these interrelationships in section 4. There is also a long history of applications of Kakeya sets to construct counterexamples in pointwise convergence questions; we will not discuss this here, but see e.g. [25] and [54].

For various reasons it is better to look also at a more quantitative formulation in terms of a maximal operator. If $\delta > 0$, $e \in S^{n-1}$, $a \in \mathbb{R}^n$ then we define

$$T_e^\delta(a) = \{x \in \mathbb{R}^n : |(x - a) \cdot e| \leq \tfrac{1}{2}, \; |(x - a)^\perp| \leq \delta\}$$

where $x^\perp = x - (x \cdot e)e$. Thus $T_e^\delta(a)$ is essentially the δ-neighborhood of the unit line segment in the e direction centered at a. If $f : \mathbb{R}^n \to \mathbb{R}$ then we define its Kakeya maximal function $f_\delta^* : S^{n-1} \to \mathbb{R}$ via

$$f_\delta^*(e) = \sup_{a \in \mathbb{R}^n} \frac{1}{|T_e^\delta(a)|} \int_{T_e^\delta(a)} |f|$$

1991 *Mathematics Subject Classification.* Primary 42B99.

This definition is due to Bourgain [7]. It is one of several similar maximal functions that have been considered, going back at least to [18].

Open question 2: Is there an estimate

$$\forall \epsilon > 0 \; \exists C_\epsilon : \|f_\delta^*\|_{L^n(S^{n-1})} \le C_\epsilon \delta^{-\epsilon} \|f\|_n \; \forall f \qquad (2)$$

Roughly speaking, this question is related to question 1 in the same way as the Hardy–Littlewood maximal theorem is related to Lebesgue's theorem on points of density. As was observed by Bourgain [7], an affirmative answer to question 2 implies an affirmative answer to question 1; see Lemma 1.6 below. Once again, when $n = 2$ it is well known that the answer to question 2 is affirmative, [18] and [7]. In higher dimensions, partial results are known paralleling the results on question 1.

Questions 1 and 2 clearly have a combinatorial side to them, and the point of view we will adopt here is to try to approach the combinatorial issues directly using ideas from the combinatorics literature. In this connection let us mention a basic principle in graph theory (the "Zarankiewicz problem"; see [5], [24], [40] for this and generalizations): fix s and suppose that $\{a_{ij}\}_{i=1}^{n}{}_{j=1}^{m}$ is an $n \times m$ $(0,1)$ matrix with no $s \times s$ submatrix of 1's. Then there is a bound

$$|\{(i,j) : a_{ij} = 1\}| \le C_s \min(mn^{1-\frac{1}{s}} + n, nm^{1-\frac{1}{s}} + m) \qquad (3)$$

To see the relationship between this sort of bound and Kakeya, just note that if $\{\ell_j\}_{j=1}^{m}$ are lines and $\{p_i\}_{i=1}^{n}$ are points, then the "incidence matrix"

$$a_{ij} = \begin{cases} 1 & \text{if } p_i \in \ell_j \\ 0 & \text{if } p_i \notin \ell_j \end{cases}$$

will contain no 2×2 submatrix of 1's, since two lines intersect in at most one point. Much of what we will say below will have to do with attempts to modify this argument, and also more sophisticated arguments in incidence geometry (e.g. [17]) to make them applicable to "continuum" problems such as Kakeya.

There are several difficulties with such an approach. It is sometimes unclear whether applying the combinatorial techniques in the continuum should be simply a matter of extra technicalities or whether new phenomena should be expected to occur, and furthermore many of the related discrete problems are regarded as being very difficult. A classical example is the Erdős unit distance problem (see [17] and [40]) and other examples will be mentioned in section 3.

Of course, much work has been done in the opposite direction, applying harmonic analysis techniques to questions of a purely geometrical appearance. A basic example is the spherical maximal theorem of Stein [51], and various Strichartz type inequalities as well as the results on the distance set problem in [22], [11] are also fairly close to the subject matter of this paper. However, we will not present any work of this nature here.

The paper is organized as follows. In section 1 we discuss the two dimensional Kakeya problem, in section 2 we discuss the higher dimensional Kakeya problem and in section 3 we discuss analogous problems for circles in the plane. Finally in section 4 we discuss the Fefferman construction and a related construction of Bourgain [9] which connects the Kakeya problem also to estimates of Dirichlet series. Section 4 contains several references to the recent literature on open problems regarding oscillatory integrals, but it is not a survey. Further references are in [10], [58], and especially [52].

We have attempted to make the presentation self-contained insofar as is possible. In particular we will present some results and arguments which are known or almost known but for which there is no easy reference.

The author is grateful for the opportunity to speak at the conference and to publish this article.

List of notation

α:: greatest integer less than or equal to α.

p': conjugate exponent to p, i.e. $p' = \frac{p}{p-1}$.

$D(x,r)$: the disc with center x and radius r.

$|E|$: Lebesgue measure or cardinality of the set E, depending on the context.

E^c: complement of E.

$\dim E$: Hausdorff dimension of E.

$H_s(E)$: s-dimensional Hausdorff content of E, i.e. $H_s(E) = \inf\left(\sum_j r_j^s : E \subset \bigcup_j D(x_j, r_j)\right)$.

$T_e^\delta(a)$: δ-tube in the e direction centered at a, as defined in the introduction. Sometimes we will also use the notation T_e^δ; this means any tube of the form $T_e^\delta(a)$ for some $a \in \mathbb{R}^n$.

$C(x,r)$: circle in $\mathbb{R}^2$ (or sphere in $\mathbb{R}^n$) with center x and radius r.

$C_\delta(x,r)$: annular region $\{y \in \mathbb{R}^n : r - \delta < |y - x| < r + \delta\}$.

$x \lesssim y$: $x \leq Cy$ for a suitable constant C.

1. The two dimensional case

We will start by proving the existence of measure zero Kakeya sets using a variant on the original construction which is quick and is easy to write out in closed form; to the author's knowledge the earliest reference for this approach is Sawyer [42]. A discussion of various other possible approaches to the construction may be found in [21].

For expository reasons, we make the following definitions.

A G-set is a compact set $E \subset \mathbb{R}^2$ which is contained in the strip $\{(x,y) : 0 \leq x \leq 1\}$, such that for any $m \in [0,1]$ there is a line segment contained in E connecting $x = 0$ to $x = 1$ with slope m, i.e.

$$\forall m \in [0,1] \; \exists b \in \mathbb{R} : mx + b \in E \; \forall x \in [0,1].$$

If $\ell = \{(x,y) : y = mx + b\}$ is a nonvertical line and $\delta > 0$, then $S_\ell^\delta \overset{\text{def}}{=} \{(x,y) : 0 \leq x \leq 1 \text{ and } |y - (mx + b)| \leq \delta\}$.

Remark 1.1 It is clear that existence of G-sets with measure zero will imply existence of Kakeya sets with measure zero. Note also that if ℓ is a line with slope m then S_ℓ^δ will contain segments connecting $x = 0$ to $x = 1$ with any given slope between $m - 2\delta$ and $m + 2\delta$.

We now describe the basic construction, which leads to the slightly weaker conclusion (Lemma 1.2) that there are G-sets with measure $< \epsilon$ for any $\epsilon > 0$. It can be understood in terms of the usual sliding triangle picture: start from a right triangle with vertices $(0,0)$, $(0,-1)$ and $(1,0)$; this is clearly a G-set. Subdivide it in N "1st stage" triangles by subdividing the vertical side in N equal intervals.

Leave the top triangle alone and slide the others upward so that their intersections with the line $x = 0$ all coincide. Next, for each of the 1st stage triangles, subdivide it in N 2nd stage triangles, leave the top triangle in each group alone and slide the $N - 1$ others upward until the intersections of the N triangles in the group with the line $x = \frac{1}{N}$ all coincide. Now repeat at abscissas $\frac{2}{N}, \frac{3}{N}, \ldots, \frac{N-1}{N}$.

Now we make this precise. Fix a large integer N and let $\mathcal{A}_N$ be all numbers in $[0, 1)$ whose base N expansion terminates after N digits, i.e.

$$a \in \mathcal{A}_N \iff a = \sum_{j=1}^{N} \frac{a_j}{N^j} \text{ with } a_j \in \{0, 1, \ldots N - 1\}.$$

To each $a \in \mathcal{A}_N$ we associate the line segment ℓ_a connecting the y axis to the line $x = 1$ with slope a and y intercept $-\sum_{j=1}^{N} \frac{(j-1)a_j}{N^{j+1}}$. Thus

$$\ell_a = \{(t, \phi_a(t)) : 0 \leq t \leq 1\}, \text{ where } \phi_a(t) = \sum_{j=1}^{N} \frac{(Nt - j + 1)a_j}{N^{j+1}}.$$

<u>Lemma 1.1</u> For each $t \in [0, 1]$ there are an integer $k \in \{1, \ldots, N\}$ and a set of N^{k-1} intervals each of length $2N^{-k}$, whose union contains the set $\{\phi_a(t) : a \in \mathcal{A}_N\}$.

<u>Proof</u> Choose k so that $\frac{k-1}{N} \leq t \leq \frac{k}{N}$. Define $a, b \in \mathcal{A}_N$ to be equivalent if $a_j = b_j$ when $j \leq k - 1$. There are N^{k-1} equivalence classes, and if a and b are equivalent then

$$|\phi_a(t) - \phi_b(t)| = \left| \sum_{j \geq k} \frac{(Nt - j + 1)(a_j - b_j)}{N^{j+1}} \right| \leq \sum_{j \geq k} \frac{\max(j - k, 1)|a_j - b_j|}{N^{j+1}}$$

$$\leq \frac{N - 1}{N^{k+1}} \sum_{j \geq k} \frac{\max(j - k, 1)}{N^{j-k}} \leq 2 \frac{N - 1}{N^{k+1}}$$

$$< \frac{2}{N^k} \text{ when } N \text{ is large.} \qquad \square$$

<u>Lemma 1.2</u> Let N be sufficiently large. Then there is a G-set $E \subset [0, 1] \times [-1, 1]$ which intersects every vertical line in measure $\leq \frac{4}{N}$, in particular $|E| \leq \frac{4}{N}$.

<u>Proof</u> We let

$$E_N = \bigcup_{a \in \mathcal{A}_N} S_{\ell_a}^{N^{-N}}.$$

Then E_N contains segments with all slopes between 0 and 1, by Remark 1.1. If $t \in [0, 1]$, then by Lemma 1.1 there is $k \in \{1, \ldots, N\}$ such that the intersection of E with the line $x = t$ is contained in the union of N^{k-1} intervals of length $2N^{-k} + 2N^{-N} \leq 4N^{-k}$. The lemma follows. $\qquad \square$

Existence of measure zero Kakeya sets now follows by a standard limiting argument, most easily carried out via the following lemma.

<u>Lemma 1.3</u> For every G-set E and every $\epsilon > 0, \eta > 0$, there is another G-set F, which is contained in the ϵ-neighborhood of E and has measure $< \eta$.

<u>Proof</u> Let δ be small, let $\{m_j\} = \{j\delta\}_{j=0}^{[\frac{1}{\delta}]}$ and for each j, fix a line segment $\ell_j = \{(x, y) : 0 \leq x \leq 1, y = m_j x + b_j\} \subset E$ with slope m_j connecting $x = 0$ to $x = 1$ and form the parallelogram $S_{\ell_j}^{\delta}$. Let A_j be the affine map from $[0, 1] \times [-1, 1]$ on $S_{\ell_j}^{\delta}$, $A_j(x, y) = (x, m_j x + b_j + \delta y)$ and consider $F = \bigcup_m A_m(E_N)$ for a large

enough N; here E_N is as in Lemma 1.2. A_j maps segments with slope μ to segments with slope $m + \delta\mu$ so F is a G-set. Clearly it is contained in the δ-neighborhood of E. Furthermore A_j contracts areas by a factor δ so $|A_j(E_N)| \leq 4\frac{\delta}{N}$ for each j, hence $|F| \lesssim \frac{1}{N}$. $\qquad\square$

Corollary 1.4 There are Kakeya sets with measure zero.

Proof We construct a sequence $\{F_n\}_{n=0}^\infty$ of G-sets, and a sequence of numbers $\{\epsilon_n\}_{n=0}^\infty$ converging to zero such that the following properties hold when $n \geq 1$; here $F(\epsilon) \overset{\text{def}}{=} \{x : \mathrm{dist}(x, F) < \epsilon\}$ is the ϵ-neighborhood of F and $\overline{E}$ is the closure of E.

(i) $F_n(\epsilon_n) \subset F_{n-1}(\epsilon_{n-1})$.

(ii) $|\overline{F_n(\epsilon_n)}| < 2^{-n}$.

Namely, we take F_0 to be any G-set, and we set $\epsilon_0 = 1$. If $n \geq 1$ and if F_{n-1} and ϵ_{n-1} have been constructed then we obtain F_n by applying Lemma 1.3 with $\epsilon = \epsilon_{n-1}$ and $\eta = 2^{-n}$. Since F_n is compact, (i) and (ii) will then hold provided ϵ_n is sufficiently small.

The set $\bigcap_n \overline{F_n(\epsilon_n)}$ is then a G-set with measure zero. $\qquad\square$

Remarks 1.2. The construction above easily gives the following variant (used e.g. in [23]): with $\delta = \frac{1}{10}N^{-N}$, there is a family of disjoint δ-tubes

$$\{T_{e_j}^\delta(x_j)\}_{j=1}^M \subset \mathbb{R}^2$$

where $M \approx \delta^{-1}$ with the property that the union of the translated tubes $T_{e_j}^\delta(x_j + 2e_j)$ has measure $\lesssim \frac{1}{N}$.

Namely, a calculation shows that if $a, b \in \mathcal{A}$ and $a < b$ then $\phi_a(1) < \phi_b(1)$, i.e. the ordering of the intersection points between the ℓ_a and the line $x = 1$ is the same as the ordering of slopes. Hence if we regard ℓ_a as extended to a complete line, then no two ℓ_a's intersect in the region $x > 1$, and in fact in the region $x > 2$ any two of them are at least N^{-N} apart. Now for each $a \in \mathcal{A}_N$ we form the rectangle R_a with length 1, width $\frac{1}{5}N^{-N}$, axis along the line ℓ_a and bottom right corner on the line $x = 1$. Clearly $R_a \subset S_a$, so $\bigcup_a R_a$ is small by Lemma 1.2. On the other hand, if R_a is translated to the right along its axis by distance 2 then the resulting rectangles are disjoint. We may therefore take $\{T_{e_j}^\delta(x_j)\}$ to be the set of translated rectangles.

1.3. Analogous statements in higher dimensions may be obtained using dummy variables.

Measure zero Kakeya sets in $\mathbb{R}^n$ may be constructed by taking the product of a Kakeya set in $\mathbb{R}^2$ with a closed disc of radius $\frac{1}{2}$ in $\mathbb{R}^{n-2}$ (or for that matter with any Kakeya set in $\mathbb{R}^{n-2}$), and a family of roughly $\delta^{-(n-1)}$ disjoint $T_e^\delta(a)$'s such that the union of the tubes $T_e^\delta(a + 2e)$ has small measure may be obtained by taking the products of the tubes in Remark 1.2 with a family of $\delta^{-(n-2)}$ disjoint δ-discs in $\mathbb{R}^{n-2}$.

We now discuss the positive results on questions 1 and 2 in dimension two. Proposition 1.5 was first stated and proved in [7] although a similar result for a related maximal function was proved earlier in [18].

We will work with restricted weak type estimates instead of with L^p estimates; this is known to be equivalent except for the form of the $\delta^{-\epsilon}$ terms.[1] We will say (see e.g. [**53**]) that an operator T has restricted weak type norm $\leq A$, written

$$\|Tf\|_{q,\infty} \leq A\|f\|_{p,1}$$

if $\left|\{x : |T\chi_E(x)| \geq \lambda\}\right| \leq \left(\frac{A|E|^{\frac{1}{p}}}{\lambda}\right)^q$ for all sets E with finite measure and all $\lambda \in (0,1]$; here χ_E is the characteristic function of E.

Proposition 1.5 The restricted weak type $(2,2)$ norm of the Kakeya maximal operator $f \to f_\delta^*$ in $\mathbb{R}^2$ is $\lesssim (\log \frac{1}{\delta})^{\frac{1}{2}}$.

More explicitly, suppose that $E \subset \mathbb{R}^2$ and $\lambda \in (0,1]$. Let $f = \chi_E$, and let $\Omega = \{e \in S^1 : f_\delta^*(e) \geq \lambda\}$. Then

$$|\Omega| \lesssim \log \frac{1}{\delta} \frac{|E|}{\lambda^2}.$$

<u>Proof</u> Let $\theta(e,f)$ be the unoriented angle subtended by the directions e and f, i.e. $\theta(e,f) = \arccos(e \cdot f)$. We start by mentioning two trivial but important facts. First, in $\mathbb{R}^n$, the intersection of the tubes $T_e^\delta(a)$ and $T_f^\delta(b)$ satisfies

$$(4) \qquad\qquad \mathrm{diam}(T_e^\delta(a) \cap T_f^\delta(b)) \lesssim \frac{\delta}{\theta(e,f) + \delta}$$

for any a and b and therefore also

$$(5) \qquad\qquad |T_e^\delta(a) \cap T_f^\delta(b)| \lesssim \frac{\delta^n}{\theta(e,f) + \delta}.$$

Next, if Ω is a set on the unit sphere $S^{n-1} \subset \mathbb{R}^n$ and if $\delta > 0$ then the δ-entropy $\mathcal{N}_\delta(\Omega)$ (maximum possible cardinality M for a δ-separated subset $\{e_j\}_{j=1}^M \subset \Omega$) satisfies

$$(6) \qquad\qquad \mathcal{N}_\delta(\Omega) \gtrsim \frac{|\Omega|}{\delta^{n-1}}.$$

Now we assume $n = 2$ and give the proof of the proposition. Fix a δ-separated $\{e_j\}_{j=1}^M \subset \Omega$, where $M \gtrsim \frac{|\Omega|}{\delta}$. For each j, there is a tube $T_j = T_{e_j}^\delta(a_j)$ with $|T_j \cap E| \geq \lambda|T_j| \approx \lambda\delta$. Thus

$$M\lambda\delta \lesssim \sum_j |T_j \cap E| = \int_E \sum_j \chi_{T_j} \leq |E|^{\frac{1}{2}} \left\|\sum_j \chi_{T_j}\right\|_2$$

$$= |E|^{\frac{1}{2}} \left(\sum_{j,k} |T_j \cap T_k|\right)^{\frac{1}{2}} \lesssim |E|^{\frac{1}{2}} \left(\sum_{j,k} \frac{\delta^2}{\theta(e_j,e_k) + \delta}\right)^{\frac{1}{2}}.$$

For fixed k the sum over j is $\lesssim \sum_{j:|j-k|\leq \frac{C}{\delta}} \frac{\delta^2}{|j-k|\delta+\delta} \lesssim \delta \log \frac{1}{\delta}$. We conclude that $M\lambda\delta \lesssim |E|^{\frac{1}{2}}(M\delta \log \frac{1}{\delta})^{\frac{1}{2}}$ which gives the result since $M \gtrsim \frac{|\Omega|}{\delta}$.   □

Now we show how to pass to the Hausdorff dimension statement. The next result is Lemma 2.15 in [**7**].

---

[1]We work with restricted weak type estimates for expository reasons only. We believe this makes the results more transparent; however, it is well known that actually $\|f_\delta^*\|_{L^2(S^1)} \lesssim (\log \frac{1}{\delta})^{\frac{1}{2}}\|f\|_2$. The latter estimate is proved in [**7**] and also follows from the proof below, plus duality, as in [**18**].

Lemma 1.6 Assume an estimate in $\mathbb{R}^n$

(7) $$\|f_\delta^*\|_{q,\infty} \leq C\delta^{-\alpha}\|f\|_{p,1}.$$

Then Kakeya sets have dimension at least $n - p\alpha$.

Proof Fix $s < n - p\alpha$. Let E be a Kakeya set and for each $e \in S^{n-1}$, fix a point x_e such that $x_e + te \in E$ when $t \in [-\frac{1}{2}, \frac{1}{2}]$. We have to bound $H_s(E)$ from below, so fix a covering of E by discs $D_j = D(x_j, r_j)$. We can evidently suppose all r_j's are ≤ 1.

Let $\Sigma_k = \{j : 2^{-k} \leq r_j \leq 2^{-(k-1)}\}$, $\nu_k = |\Sigma_k|$ and $E_k = E \cap (\bigcup\{D_j : j \in \Sigma_k\})$. Also let $\tilde{D}_j = D(x_j, 2r_j)$, and $\tilde{E}_k = \bigcup\{\tilde{D}_j : j \in \Sigma_k\}$.

Then $\bigcup_k E_k = E$, so for each e the pigeonhole principle implies

$$\left|\{t \in [-\tfrac{1}{2}, \tfrac{1}{2}] : x_e + te \in E_k\}\right| \geq \frac{c}{k^2}$$

for some $k = k_e$, where $c = \frac{6}{\pi^2}$. By the pigeonhole principle again, we can find a fixed k so that $k = k_e$ when $e \in \Omega$, where $\Omega \subset S^{n-1}$ has measure $\geq \frac{c}{k^2}$. With this k, we note that $\tilde{E}_k$ contains a disc of radius 2^{-k} centered at each point of E_k; it follows easily that if $e \in \Omega$ then $\left|T_e^{2^{-k}}(x_e) \cap \tilde{E}_k\right| \gtrsim k^{-2}\left|T_e^{2^{-k}}(x_e)\right|$. With $f = \chi_{\tilde{E}_k}$ we therefore have

$$\left|\{e : f_{2^{-k}}^*(e) \geq C^{-1}k^{-2}\}\right| \gtrsim k^{-2}.$$

On the other hand, by the assumption (7)

$$\left|\{e : f_{2^{-k}}^*(e) \geq C^{-1}k^{-2}\}\right| \lesssim \left(k^2 2^{k\alpha}|\tilde{E}_k|^{\frac{1}{p}}\right)^q$$

and $|\tilde{E}_k| \lesssim \nu_k 2^{-kn}$. So $\left(k^2 2^{k\alpha}(\nu_k 2^{-kn})^{\frac{1}{p}}\right)^q \gtrsim k^{-2}$, or equivalently

$$\nu_k \gtrsim k^{-\frac{2}{p}(1+\frac{1}{q})} 2^{k(n-p\alpha)}.$$

Letting $\epsilon = n - p\alpha - s > 0$, we have $\sum_j r_j^s \gtrsim \nu_k 2^{-ks} \gtrsim k^{-2p(1+\frac{1}{q})} 2^{k\epsilon} \geq$ constant. $\square$

Applying this with $p = n = 2$ we see that Proposition 1.5 implies Davies' theorem that Kakeya sets in $\mathbb{R}^2$ have dimension 2. Likewise it follows that yes on question 2 for a given n will imply yes on question 1 for the same n.

Remark 1.4 It is clear that the logarithmic factor in Proposition 1.5 cannot be dropped entirely, since then the above argument would show that measure zero Kakeya sets could not exist. In fact it has been known for a long time that the exponent $\frac{1}{2}$ cannot be improved, and U. Keich [29] recently showed that even a higher order improvement is not possible in Proposition 1.5 or in its corollary on L^p for $p > 2$. On the other hand, a number of related questions concerning logarithmic factors have been solved only recently or are still open. In particular we should mention the results of Barrionuevo [2] and Katz [26], [27] on the question of maximal functions defined using families of directions in the plane.

Remark 1.5 An interesting open question in $\mathbb{R}^2$ is the following one, which arose from work of Furstenburg.

For a given $\alpha \in (0, 1]$, suppose that E is a compact set in the plane, and for each $e \in S^1$ there is a line ℓ_e with direction e such that $\dim(\ell_e \cap E) \geq \alpha$. Then what is the smallest possible value for $\dim E$?

Easy results here are that $\dim E \geq \max(2\alpha, \frac{1}{2}+\alpha)$ and that there is an example with $\dim E = \frac{1}{2} + \frac{3}{2}\alpha$. We give proofs below. Several people have unpublished results on this question and it is unlikely that the author was the first to observe these bounds; in all probability they are due to Furstenburg and Katznelson.

The analogous discrete question is solved by the following result due to Szemeredi and Trotter [56] (see also [17], [40], [55]).

Suppose we are given n points $\{p_i\}$ and k lines $\{\ell_j\}$ in the plane. Define a line and point to be <u>incident</u>, $p \sim \ell$, if p lies on ℓ. Let $\mathcal{I} = \{(i,j) : p_i \sim \ell_j\}$. Then $|\mathcal{I}| \lesssim (kn)^{\frac{2}{3}} + k + n$, and this bound is sharp.

We note that the weaker bound $|\mathcal{I}| \lesssim (kn)^{\frac{3}{4}} + k + n$ follows from (3) and was known long before [56]. To see the analogy with the Hausdorff dimension question, reformulate the Szemeredi–Trotter bound as follows: if each line is incident to at least μ points ($\mu \gg 1$), then (since $|\mathcal{I}| \geq k\mu$)

$$(8) \qquad\qquad n \gtrsim \min(\mu^{\frac{3}{2}}k^{\frac{1}{2}}, \mu k).$$

Now assume say [2] that E has a covering by n discs D_i of radius δ. Consider a set of $k \approx \delta^{-1}$ δ-separated directions $\{e_j\}$. For each j the line ℓ_{e_j} will intersect D_i for at least $\delta^{-\alpha}$ values of i. We now pretend that we can replace points by the discs D_i in Szemeredi–Trotter and apply (8) with $\mu = \delta^{-\alpha}$, $k = \delta^{-1}$. Since $k \geq \mu$ we would obtain $n \gtrsim \delta^{-\frac{1}{2}-\frac{3}{2}\alpha}$, i.e. that the bound $\dim E \geq \frac{1}{2} + \frac{3}{2}\alpha$ should hold.

In one situation to be discussed in section 3, it turns out that this kind of heuristic argument can be justified leading to a theorem in the continuum. In other situations such as the present one, it seems entirely unclear whether this should be the case or not, but still the discrete results suggest plausible conjectures.

If correct the bound $\dim E \geq \frac{1}{2} + \frac{3}{2}\alpha$ would be best possible by essentially the same example (due to Erdős, see [40]) that shows the Szemeredi–Trotter bound is sharp.

We start by recalling that if $\{n_j\}$ is a sequence of integers which increases sufficiently rapidly, and if $\alpha \in (0,1)$ then the set

$$T \stackrel{\mathrm{def}}{=} \left\{ x \in (\tfrac{1}{4}, \tfrac{3}{4}) : \forall j \; \exists p, q \in \mathbb{Z} : q \leq n_j^\alpha \text{ and } |x - \tfrac{p}{q}| \leq n_j^{-2} \right\}$$

has Hausdorff dimension α. This is a version of Jarnik's theorem—see [21, p. 134, Theorem 8.16(b)].

It follows that the set

$$T' = \left\{ t : \frac{1-t}{t\sqrt{2}} \in T \right\}$$

also has dimension α.

For fixed n, consider the set of all line segments ℓ_{jk} connecting a point $(0, \frac{i}{n})$ to a point $(1, \frac{k}{n}\sqrt{2})$, where j and k are any integers between 0 and $n-1$. Thus $\ell_{jk} = \{(x, \phi_{jk}(x)) : 0 \leq x \leq 1\}$ where $\phi_{jk}(x) = (1-x)\frac{j}{n} + x\frac{k}{n}\sqrt{2}$. It follows using e.g. [32, p. 124, example 3.2] that every number in $[0,1]$ differs by $\lesssim n^{-2}(\log n)^2$ from the slope of one of the ℓ_{jk}'s, so the set

$$G_n \stackrel{\mathrm{def}}{=} \bigcup_{jk} S_{\ell_{jk}}^{n^{-2}(\log n)^3}$$

---

[2] In this heuristic argument we ignore the distinction between Hausdorff and Minkowski dimension.

is a G-set.
   Define

$$Q_n = \left\{ t : \frac{1-t}{t\sqrt{2}} \text{ is a rational number } \frac{p}{q} \in \left(\frac{1}{4}, \frac{3}{4}\right) \text{ with denominator } q \leq n^\alpha \right\}.$$

If $t \in Q_n$, then let $S(t) \overset{\text{def}}{=} \{\phi_{jk}(t)\}_{j,k=0}^{n-1}$. For any j and k we have

$$(t\sqrt{2})^{-1} \phi_{jk}(t) = \frac{pj + qk}{qn},$$

a rational with denominator qn. We conclude that $|S(t)| \lesssim qn \leq n^{1+\alpha}$, hence $\left| \bigcup (S(t) : t \in Q_n) \right| \lesssim n^{1+3\alpha}$ and

   (∗) The set $\{(x,y) \in G_n : |x - t| \leq \frac{1}{n^2} \text{ for some } t \in Q_n\}$ is contained in the union of $\lesssim n^{1+3\alpha}$ discs of radius $n^{-2}(\log n)^3$.

   Now we let $\{n_j\}$ increase rapidly and will recursively construct compact sets F_j such that $F_{j+1} \subset F_j$, each F_j is a G-set and the set $\{(x,y) \in F_j : x \in T'\}$ is contained in the union of $n_j^{1+3\alpha} \log n_j$ discs of radius $n_j^{-2}(\log n_j)^3$. Namely, let F_0 be any G-set. If F_j has been constructed it will be of the form

$$\bigcup_{i=1}^{M} S_{\ell_i}^\delta$$

for a certain δ, where $\ell_i = \{(x, m_i x + b_i) : 0 \leq x \leq 1\}$ for suitable m_i and b_i, and every number in $[0,1]$ is within δ of one of the m_i. As in the proof of Lemma 1.3 we let $A_i(x,y) = (x, m_i x + \delta y + b_i)$. We make n_{j+1} sufficiently large and define

$$F_{j+1} = \bigcup_{i=1}^{M} A_i(G_{n_{j+1}}).$$

Clearly $F_{j+1} \subset F_j$, and it follows as in Lemma 1.3 that the resulting set is a G-set. The covering property is also essentially obvious from (∗) provided n_{j+1} is large enough, say $\log(n_{j+1}) \gg M$.

   Let $F = \bigcap_j F_j$, and let $E = \{(x,y) \in F : x \in T'\}$. Then the covering property in the construction of F_j implies that $\dim E \leq \frac{1}{2}(1 + 3\alpha)$. On the other hand F is a G-set, and if ℓ is a line segment contained in F, then $\dim(\ell \cap E) = \dim T' \geq \alpha$. This completes the construction.

   We now discuss the bound $\dim E \geq \max(2\alpha, \frac{1}{2} + \alpha)$. The bound $\dim E \geq 2\alpha$ can be derived from Proposition 1.5 by an argument like the proof of Lemma 1.6; we will omit this. To prove the bound $\dim E \geq \frac{1}{2} + \alpha$ (which corresponds to the easy $|\mathcal{I}| \lesssim (kn)^{\frac{3}{4}} + k + n$ under the above heuristic argument) fix a compact set E and for each $e \in S^1$ a line ℓ_e which intersects E in dimension $\geq \alpha$. Let $\{D_j\} = \{D(x_j, r_j)\}$ be a covering. Fix $\beta_1 < \beta < \alpha$; we have to bound $\sum_j r_j^{\frac{1}{2} + \beta_1}$ from below. As in the proof of Lemma 1.6 we let $\Sigma_k = \{j : 2^{-k} \leq r_j \leq 2^{-(k-1)}\}$, $\nu_k = |\Sigma_k|$ and $E_k = E \cap \left(\bigcup \{D_j : j \in \Sigma_k\} \right)$. We start by choosing a number k and a subset $\Omega \subset S^1$ with measure $\gtrsim \frac{1}{k^2}$ such that if $e \in \Omega$ then $H_\beta(\ell_e \cap E_k) \geq C^{-1}k^{-2}$, using the pigeonhole principle as in the proof of Lemma 1.6. Let $\gamma = \frac{2}{\beta}$. Since $H_\beta(I) \leq |I|^\beta$ for any interval I, it follows that for a suitable numerical constant C, and for any $e \in \Omega$ there are <u>two</u> intervals $I_e^\pm$ on ℓ_e which are $C^{-1}k^{-\gamma}$- separated and such that $H_\beta(E_k \cap I_e^\pm) \gtrsim k^{-2}$. Let $\{e_i\}_{i=1}^M$ be a 2^{-k}-separated subset of Ω with $M \gtrsim \frac{2^k}{k^2}$ (see (6)) and define

$$(9) \qquad \mathcal{T} = \{(j_+, j_-, i) \in \Sigma_k \times \Sigma_k \times \{1, \dots, M\} : I_{e_i}^+ \cap E_k \cap D_{j_+} \neq \varnothing, I_{e_i}^- \cap E_k \cap D_{j_-} \neq \varnothing\}.$$

We will count $\mathcal{T}$ in two different ways.

First fix j_+ and j_- and consider how many values of i there can be with $(j_+, j_-, i) \in \mathcal{T}$. We will call such a value of i <u>allowable</u>. If the distance between D_{j_+} and D_{j_-} is small compared with $k^{-\gamma}$ then there is no allowable i, since the distance between $I_{e_i}^+$ and $I_{e_i}^-$ is always $\geq C^{-1}k^{-\gamma}$. On the other hand if the distance between D_{j_+} and D_{j_-} is $\gtrsim k^{-\gamma}$, then because the $\{e_i\}$ are 2^{-k}-separated, it follows that there are $\lesssim k^\gamma$ i's such that ℓ_{e_i} intersects both D_{j_+} and D_{j_-}. Hence in either case there are $\lesssim k^\gamma$ allowable i's. Summing over j_+ and j_- we conclude that

$$(10) \qquad |\mathcal{T}| \lesssim k^\gamma \nu_k^2.$$

On the other hand, for any fixed i, the lower bound $H_\beta(E_k \cap I_{e_i}^+) \gtrsim k^{-2}$ implies there are $\gtrsim k^{-\gamma}2^{k\beta}$ values of j_+ such that $I_{e_i}^+ \cap E_k \cap D_{j_+} \neq \varnothing$ and similarly with $+$ replaced by $-$. So $|\mathcal{T}| \gtrsim M(k^{-\gamma}2^{k\beta})^2$. Comparing this bound with (10) we conclude that

$$\nu_k \gtrsim k^{-\frac{3}{2}\gamma}2^{k\beta}\sqrt{M} \gtrsim k^{-(1+\frac{3}{2}\gamma)}2^{(\frac{1}{2}+\beta)k} \gtrsim 2^{(\frac{1}{2}+\beta_1)k}$$

and therefore $\sum_{j \in \Sigma_k} r_j^{\frac{1}{2}+\beta_1} \geq \text{constant}.$ $\qquad \square$

2. The higher dimensional case

We will first make a few remarks about the corresponding problem over finite fields, which is the following:

Let $\mathbb{F}_q$ be the field with q elements and let V be an n-dimensional vector space over $\mathbb{F}_q$. Let E be a subset of V which contains a line in every direction, i.e.

$$\forall e \in V\backslash\{0\} \exists a \in V : a + te \in E \ \forall t \in \mathbb{F}_q.$$

Does it follow that $|E| \geq C_n^{-1}q^n$?

Of course C_n should be independent of q. One could ask instead for a bound like $\forall \epsilon > 0 \ \exists C_{n\epsilon} : |E| \geq C_{n\epsilon}^{-1}q^{n-\epsilon}$ or could restrict to the case of prime fields $\mathbb{F}_p$ or fields with bounded degree over the prime field.

So far as I have been able to find out this question has not been considered, and the simple result below corresponds to what is known in the Euclidean case.

Proposition 2.1 In the above situation $|E| \geq C_n^{-1}q^{\frac{n+2}{2}}$.

We give the proof since it is based on the same idea as the $\mathbb{R}^n$ proof but involves no technicalities.

First consider the case $n = 2$, which is analogous to Proposition 1.5. We will actually prove the following more general statement, which we need below: suppose (with $\dim V = 2$) that E contains at least $\frac{q}{2}$ points on a line in each of m different directions. Then

$$(11) \qquad |E| \gtrsim mq.$$

To prove (11), let $\{\ell_j\}_{j=1}^m$ be the lines. Any two distinct ℓ_j's intersect in a point. Accordingly

$$\tfrac{1}{2}qm \le \sum_j |E \cap \ell_j| \le |E|^{\frac{1}{2}} \left(\sum_{jk} |\ell_j \cap \ell_k| \right)^{\frac{1}{2}} = |E|^{\frac{1}{2}} (m(m-1+q))^{\frac{1}{2}} \le |E|^{\frac{1}{2}} (mq)^{\frac{1}{2}}$$

where we used that $m \le q+1$. It follows that $|E| \gtrsim mq$. Taking $m = q+1$ we obtain the two dimensional case of Proposition 2.1.

Now assume $n \ge 3$. Then E contains $\frac{q^n-1}{q-1} \approx q^{n-1}$ lines $\{\ell_j\}$. Fix a number μ and define a <u>high multiplicity line</u> to be a line ℓ_k with the following property: for at least $\frac{q}{2}$ of the q points $x \in \ell_k$, the set $\{j : x \in \ell_j\}$ has cardinality at least $\mu + 1$. Consider two cases: (i) no high multiplicity line exists (ii) a high multiplicity line exists.

In case (i) we define $\tilde{E} = \{x \in E : x \text{ belongs to } \le \mu \; \ell_j\text{'s}\}$. Then $\tilde{E}$ intersects each ℓ_j in at least $\frac{q}{2}$ points, by definition of case (i). Each point of $\tilde{E}$ belongs to at most μ ℓ_j's so we may conclude that

$$|E| \ge |\tilde{E}| \ge \mu^{-1} \sum_j |\tilde{E} \cap \ell_j| \gtrsim \mu^{-1} q \cdot q^{n-1}.$$

In case (ii), let $\{\Pi_i\}$ be an enumeration of the 2- planes containing ℓ_k. By definition of high multiplicity line there are at least $\frac{\mu q}{2}$ lines ℓ_j, $j \ne k$, which intersect ℓ_k. Each one of them is contained in a unique Π_i, and contains $q-1$ points of Π_i which do not lie on ℓ_k. Let $\mathcal{L}_i$ be the set of lines ℓ_j which are contained in a given Π_i. Then by (11) we have $\left| E \cap \Pi_i \cap (V \backslash \ell_k) \right| \gtrsim q|\mathcal{L}_i|$. The sets $\Pi_i \cap (V \backslash \ell_k)$ are pairwise disjoint so we can sum over i to get $|E| \gtrsim q \sum_i |\mathcal{L}_i| \ge \frac{q^2 \mu}{2}$.

If we take μ to be roughly $q^{\frac{n-2}{2}}$ we obtain $|E| \gtrsim q^{\frac{n+2}{2}}$ in either case (i) or (ii), hence the result. $\qquad \square$

<u>Remark 2.1</u> General finite fields do not always resemble the Euclidean case in this sort of problem. For example, the Szemeredi–Trotter theorem is easily seen to be false (e.g. [**5**, p. 75]). A counterexample involving one line in each direction as in remark 1.5 may be obtained in the following way: let $q = p^2$ with p prime, let α be a generator of $\mathbb{F}_q$ over $\mathbb{F}_p$ and in the two dimensional vector space V over $\mathbb{F}_q$, let ℓ_{jk} be the line connecting $(0,j)$ to $(1, k\alpha)$. Here j and k are in $\mathbb{F}_p$. This is a set of p^2 lines containing one line in each direction other than the vertical. For given $t \in \mathbb{F}_q$, let $S_t = \{y \in \mathbb{F}_q : (t,y) \in \bigcup_{jk} \ell_{jk}\}$. If t is such that $\alpha \frac{t}{1-t} \in \mathbb{F}_p$ then it is easily seen that S_t coincides with $(1-t)\mathcal{F}_p$, and if $t = 1$ then $S_t = \alpha\mathbb{F}_p$. This gives p "bad" values of t such that S_t has cardinality p. Let $E = \bigcup_t \{(t,y) : y \in S_t\}$, where the union is taken over the bad values of t. Then $\{\ell_{jk}\}$ and E give a configuration of p^2 lines and p^2 points with p^3 incidences, matching the trivial upper bound from (3).

In the $\mathbb{R}^n$ context, arguments like the proof of Proposition 2.1 can still be used, except that one has to work with tubes instead of lines and measure instead of cardinality, and take into account such issues as that the size of the intersection of two tubes will depend on the angle of intersection via (4). This was perhaps first done by Cordoba (e.g. [**18**]—see the proof of Proposition 1.5 above). We will present here the "bush" argument from [**7**, p. 153-4], which shows the following:

<u>Proposition 2.2</u> $\|f_\delta^*\|_{n+1,\infty} \le C_n \delta^{-\frac{n-1}{n+1}} \|f\|_{\frac{n+1}{2},1}$.

<u>Proof</u> Using (6), we see that what must be shown is the following: if $\{T^\delta_{e_j}\}^M_{j=1}$ are tubes with δ-separated directions, E is a set and $|E \cap T_{e_j}| \geq \lambda |T^\delta_{e_j}|$, then

$$(12) \qquad\qquad |E| \gtrsim \delta^{\frac{n-1}{2}} \lambda^{\frac{n+1}{2}} \sqrt{M}.$$

To this end we fix a number μ ("multiplicity") and consider the following two possibilities:

(i) (low multiplicity) No point of E belongs to more than μ tubes $T^\delta_{e_j}$.
(ii) (high multiplicity) Some point $a \in E$ belongs to more than μ tubes $T^\delta_{e_j}$.

In case (i) it is clear that $|E| \gtrsim \mu^{-1} \sum_j |E \cap T^\delta_{e_j}|$, hence

$$(13) \qquad\qquad |E| \gtrsim \mu^{-1} M \lambda \delta^{n-1}.$$

In case (ii) we fix a point a as indicated and may assume that a belongs to $T^\delta_{e_j}$ when $j \leq \mu + 1$. If C_0 is a suitably large fixed constant, then $\left|T^\delta_{e_j} \cap D(a, C_0^{-1}\lambda)\right| \leq \frac{1}{2}|T^\delta_{e_j}|$. Accordingly, for $j \leq \mu + 1$, we have

$$|E \cap T^\delta_{e_j} \cap D(a, C_0^{-1}\lambda)^c| \geq \frac{\lambda}{2}|T^\delta_{e_j}| \gtrsim \lambda \delta^{n-1}.$$

If $j, k \leq \mu$ then $T^\delta_{e_j} \cap T^\delta_{e_k}$ contains a and has diameter $\lesssim \frac{\delta}{\theta(e_j, e_k)}$ by (4). It follows that if $\theta(e_j, e_k) \geq C_1 \frac{\delta}{\lambda}$ for a suitably large C_1, then the sets $E \cap T^\delta_{e_j} \cap D(a, C_0^{-1}\lambda)^c$ and $E \cap T^\delta_{e_k} \cap D(a, C_0^{-1}\lambda)^c$ are disjoint. We conclude that

$$|E| \gtrsim \mathcal{N} \cdot \lambda \delta^{n-1}$$

where $\mathcal{N}$ is the maximum possible cardinality for a $C_1 \frac{\delta}{\lambda}$-separated subset of $\{e_j\}^{\mu+1}_{j=1}$. Since the $\{e_j\}$ are δ- separated, we have $\mathcal{N} \gtrsim \lambda^{n-1}\mu$ and therefore

$$(14) \qquad\qquad |E| \gtrsim \lambda^n \delta^{n-1}\mu.$$

We conclude that for any given μ either (13) or (14) must hold. Taking $\mu \approx \lambda^{-(\frac{n-1}{2})}\sqrt{M}$ we get (12). $\qquad\qquad\qquad\qquad\qquad\qquad\qquad\qquad\qquad\qquad\square$

<u>Further remarks</u> 2.2. Bourgain [7] also gave an additional argument leading to an improved result which implies $\dim(\text{Kakeya}) \geq \frac{n+1}{2} + \epsilon_n$, where ϵ_n is given by a certain inductive formula (in particular $\epsilon_3 = \frac{1}{3}$). A more efficient argument was then given by the author [60], based on considering families of tubes which intersect a line instead of a point as in the bush argument; this is the continuum analogue of the proof of Proposition 2.1. It gives the bound

$$(15) \qquad\qquad \forall \epsilon \; \exists C_\epsilon : \|f^*_\delta\|_q \leq C_\epsilon \delta^{-(\frac{n}{p}-1)-\epsilon}\|f\|_p,$$

where $p = \frac{n+2}{2}$ and $q = (n-1)p'$. This is the estimate on L^p which would follow by interpolation with the trivial $\|f^*_\delta\|_\infty \lesssim \delta^{-(n-1)}\|f\|_1$ if the bound (2) could be proved. In particular, it implies the dimension of Kakeya sets is $\geq \frac{n+2}{2}$. Other proofs of estimates like (15) have also recently been given by Katz [28] and Schlag [45]. However in every dimension $n \geq 3$ it is unknown whether (15) holds for any $p > \frac{n+2}{2}$ and whether $\dim(\text{Kakeya}) > \frac{n+2}{2}$.

2.3. Proposition 2.2 is also a corollary of the $L^{\frac{n+1}{2}} \to L^{n+1}$ estimate for the x-ray transform due to Drury and Christ [20], [18] (see also [39], [16] for related results). Conversely, a refinement of the argument which proves (15) can be used to prove the estimate on $L^{\frac{n+2}{2}}$ which would follow from (2) and the result of [20] by interpolation, at least in the three dimensional case. See [62].

2.4. We briefly discuss some other related problems. The classical problem of Nikodym sets has been shown to be formally equivalent to the Kakeya problem by Tao [58]; we refer to his paper for further discussion. Another classical problem is the problem of $(n, 2)$ sets: suppose that E is a set in $\mathbb{R}^n$ which contains a translate of every 2-plane. Does it follow that E has positive measure? At present this is known only when $n = 3$ [33] or $n = 4$ [7]. The argument in [7], section 4 shows the following: suppose that (2) can be proved in dimension $n-1$, or more precisely that a slightly weaker result can be proved, namely that for some p and q there is an estimate

$$(16) \qquad \|f_\delta^*\|_{L^q(S^{n-2})} \lesssim \delta^{-\alpha} \|f\|_{L^p(\mathbb{R}^{n-1})} \text{ with } \alpha < \frac{1}{p}.$$

Then $(n, 2)$-sets have positive measure.

However, note that (16) would imply by Lemma 1.6 that Kakeya sets in $\mathbb{R}^{n-1}$ have dimension $\geq n-2$. In fact if an estimate (16) is true for every n then one could answer question 1 affirmatively by an argument based on the fact that the direct product of Kakeya sets is Kakeya. It may therefore be unlikely that the $(n, 2)$-sets problem can be solved without a full understanding of the Kakeya problem. However, the most recent results on it are those of [1].

2.5. If one considers curves instead of lines, then it is known that much less can be expected to be true. This first results in this direction are in [8]; see also [10], [35] and [49].

Added in proof. Bourgain recently improved over the results discussed here in sufficiently high dimensions. Specifically, for certain explicit numbers $\alpha \geq \frac{1}{2}$ the Hausdorff dimension of a Kakeya set in $\mathbb{R}^n$ is at least $\alpha(n-1) + 1$, and a similar statement is true in the setting of Proposition 2.1. We refer to his forthcoming paper for the details.

3. Circles

In this section we will discuss some analogous problems about circles in the plane, or (essentially equivalent) fine estimates for the wave equation in $2 + 1$ dimensions. These problems are much better understood than the Kakeya problem and yet they present some of the same difficulties.

A prototype result due to Bourgain [6] and Marstrand [34] independently is that

(*) A set in $\mathbb{R}^2$ which contains circles with arbitrary centers must have positive measure.

Bourgain proved a stronger result which has the same relation to (*) as question 2 does to question 1. Namely, define a maximal function

$$\mathcal{M}f(x) = \sup_r \int \left| f(x + re^{i\theta}) \right| \frac{d\theta}{2\pi}.$$

Then $\|\mathcal{M}f\|_{L^p(\mathbb{R}^2)} \lesssim \|f\|_{L^p(\mathbb{R}^2)}$, $p > 2$. As is well-known, this maximal function was introduced by Stein [51] and he proved the analogous inequality in dimensions $n \geq 3$; the range of p is then $p > \frac{n}{n-1}$. Stein's proof was based partly on the Plancherel theorem and Bourgain's argument in the two dimensional case also used the Plancherel theorem, whereas Marstrand's argument was purely geometric. We will discuss some further developments of the latter approach.

A variant on the Kakeya construction due to Besicovitch–Radó [4] and Kinney [30] shows the following:

(**) There are compact sets in the plane with measure zero containing circles of every radius between 1 and 2.

We will call such sets BRK sets. The distinction between (*) and (**) can be understood in terms of parameter counting: a set as in (**) is a subset of a 2-dimensional space containing a 1-parameter family of 1-dimensional objects, so whether it has positive measure or not can be expected to be a borderline question. This is analogous to the question of Kakeya sets which also contain $n-1$-parameter families of 1-dimensional objects. On the other hand a set as in (*) contains a 2-parameter family of 1-dimensional objects in a 2-dimensional space.

A further related remark is that analogous constructions with other 1-parameter families of circles have been done by Talagrand [57]. For example, he shows that for any smooth curve γ there are sets of measure zero containing circles centered at all points of γ.

It is natural to ask whether the dimension of a BRK set must be 2 or not. This question also has a maximal function version; the relevant maximal function is the following M_δ: if $f : \mathbb{R}^2 \to \mathbb{R}$ then $M_\delta f : [\frac{1}{2}, 2] \to \mathbb{R}$,

$$(17) \qquad M_\delta f(r) = \sup_x \frac{1}{|C_\delta(x,r)|} \int_{C_\delta(x,r)} |f|.$$

One shows analogously to Lemma 1.6 that a bound (for some $p < \infty$)

$$(18) \qquad \forall \epsilon \ \exists C_\epsilon : \|M_\delta f\|_{L^p([\frac{1}{2},2])} \leq C_\epsilon \delta^{-\epsilon} \|f\|_p$$

will imply that BRK sets have dimension 2. Note that existence of measure zero BRK sets implies the $\delta^{-\epsilon}$ factor is needed. This is similar to the situation with the two dimensional Kakeya problem. However in contrast to the latter problem it is not possible to take $p = 2$ in (18). In fact p must be at least 3; this is seen by considering the standard example $f = $ indicator function of a rectangle with dimensions $\delta \times \sqrt{\delta}$.

Remark 3.1 Sets in $\mathbb{R}^n$ with measure zero containing spheres of all radii may be shown to exist for $n \geq 3$ also, and the maximal function (17) may be defined in $\mathbb{R}^n$. However, in that case the questions mentioned above are essentially trivial, since the correct estimate for the maximal function is an $L^2 \to L^2$ estimate, is easy and implies that sets containing spheres with all radii have dimension n. Namely, the estimate

$$(19) \qquad \|M_\delta f\|_2 \lesssim \left(\log \tfrac{1}{\delta}\right)^{\frac{1}{2}} \|f\|_2$$

can be proved analogously to Proposition 1.5 and is also closely related to some of the Strichartz inequalities for the wave equation (cf. [41]), due to the fact that spherical means correspond roughly to solutions of the initial value problem $\Box u = 0$, $u(\cdot, 0) = f$, $\frac{\partial u}{\partial t}(\cdot, 0) = 0$ after taking $\frac{n-1}{2}$ derivatives. These remarks are from [31]. From a certain point of view, the "reason" why the higher dimensional case is easier is the following: if $|r - s| \approx 1$ then

(20)
$$|C_\delta(x,r) \cap C_\delta(y,s)| \approx \begin{cases} \delta^{\frac{n+1}{2}} & \text{if } C(x,r) \text{ and } C(y,s) \text{ are tangent} \\ \delta^2 & \text{if } C(x,r) \text{ and } C(y,s) \text{ are sufficiently transverse} \end{cases}$$

making the first possibility "worse" than the second in $\mathbb{R}^2$ but not in higher dimensions.

We now consider only the two dimensional case and will formulate a discrete analogy like the analogy between the Szemeredi–Trotter theorem and the question mentioned in Remark 1.5. The relevant problem in discrete geometry is

Given N circles $\{C_i\}$ in the plane, no three tangent at a point, how many pairs (i,j) can there be such that C_i is tangent to C_j?

For technical reasons we always interpret "tangent" as meaning "internally tangent", i.e. a circle $C(x,r)$ is "tangent" to $C(y,s)$, written $C(x,r)\|C(y,s)$, iff $|x - y| = |r - s|$.

We will call this the tangency counting problem. We're not aware of any literature specifically about this problem, but known techniques in incidence geometry (related to the Szemeredi–Trotter theorem) can be adapted to it without difficulty. One obtains the following bounds for $\mathcal{I} \overset{\text{def}}{=} \{(i,j) : C_i\|C_j\}$.

(i) (easy) $|\mathcal{I}| \lesssim N^{\frac{5}{3}}$. This follows from the fact that the incidence matrix

$$a_{ij} = \begin{cases} 1 & \text{if } C_i\|C_j \\ 0 & \text{otherwise} \end{cases}$$

contains no 3×3 submatrix of 1's (essentially a theorem of Apollonius: there are at most two circles which are internally tangent to three given circles at distinct points) and therefore contains at most $\mathcal{O}(N^{\frac{5}{3}})$ 1's by (3).

(ii) (more sophisticated) $\forall \epsilon > 0 \; \exists C_\epsilon < \infty : |\mathcal{I}| \lesssim N^{\frac{3}{2}+\epsilon}$. This follows readily from the techniques of Clarkson, Edelsbrunner, Guibas, Sharir and Welzl [17]. We will not discuss their work here; we just note that they prove the analogous $N^{\frac{3}{2}+\epsilon}$ bound in the three dimensional unit distance problem: in our notation, given $\{(x_i, r_i)\}_{i=1}^N \subset \mathbb{R}^2 \times \mathbb{R}$, there are $\lesssim N^{\frac{3}{2}+\epsilon}$ pairs (i,j) with $|x_i - x_j|^2 + (r_i - r_j)^2 = 1$.

There is no reason to think that the bound (ii) should be sharp.[3] However, (ii) leads to a sharp result on the BRK sets problem and a proof of the maximal inequality (18) with $p = 3$. The heuristic argument is the following: assume we know a bound $\lesssim N^\alpha$ in the tangency counting problem, where $\alpha \geq \frac{3}{2}$. Let E be a BRK set and consider its δ-neighborhood E^δ. Let $\{r_j\}_{j=1}^M$ be a maximal δ-separated subset of $[\frac{1}{2}, 2]$; then $M \approx \frac{1}{\delta}$ and E^δ contains an annulus $C_\delta(x_j, r_j)$ for each j. By (20), we should have to a first approximation $|C_\delta(x_j, r_j) \cap C_\delta(x_k, r_k)| \approx \delta^{\frac{3}{2}}$ if

---

[3]It may be more natural to consider a slightly different formulation of the problem: drop the assumption that no three circles are tangent at a point, and consider the number of points where two are tangent instead of the number of tangencies. With this reformulation, a standard example involving circles with integer center and radius shows that the exponent $\frac{4}{3}$ would be best possible as in the unit distance problem.

$C(x_j, r_j)$ and $C(x_i, r_i)$ intersect tangentially and $|C_\delta(x_j, r_j) \cap C_\delta(x_k, r_k)| \approx \delta^2$ if they intersect transversally. Accordingly we would get

$$\sum_{jk} |C_\delta(x_j, r_j) \cap C_\delta(x_k, r_k)| \lesssim \delta^{-\alpha} \cdot \delta^{\frac{3}{2}} + \delta^{-2} \cdot \delta^2 \lesssim \delta^{\frac{3}{2} - \alpha},$$

and then the argument in the proof of Proposition 1.5 shows that $|E_\delta| \gtrsim \delta^{\frac{1}{2}(\alpha - \frac{3}{2})}$, so one expects $\dim E \geq 2 - \frac{1}{2}(\alpha - \frac{3}{2})$.

It turns out that it is possible to make this argument rigorous and to obtain a corresponding result ((18) with $p = 3$) for the maximal operator. The first lemma below keeps track of the intersection of two annuli in terms of their degree of tangency; it is of course quite standard and is used in one form or another in most papers in the area, e.g. [6] and [34]. The second lemma is due to Marstrand ([34], Lemma 5.2), although he formulated it slightly differently. It gives a quantitative meaning to the theorem of Apollonius used in the proof of the $N^{\frac{5}{3}}$ tangency bound.

We introduce the following notation: if $C(x, r)$ and $C(y, s)$ are circles then

$$d((x, r), (y, s)) = |x - y| + |r - s|,$$
$$\Delta((x, r), (y, s)) = \big| |x - y| - |r - s| \big|.$$

Note that Δ vanishes precisely when the circles are "tangent." In Lemmas 3.1 and 3.2 below, we assume that all circles $C(x, r)$ etc. have centers in $D(0, \frac{1}{4})$ and radii between $\frac{1}{2}$ and 2.

<u>Lemma 3.1</u> Assume that $x \neq y$. Let $d = d((x, r), (y, s))$, $\Delta = \Delta((x, r), (y, s))$, and $e = \text{sgn}(r - s) \frac{y - x}{|y - x|}$, $\zeta = y + re$. Then

(a) $C_\delta(x, r) \cap C_\delta(y, s)$ is of measure $\lesssim \delta \cdot \frac{\delta}{\sqrt{(\delta + \Delta)(\delta + d)}}$.

(b) $C_\delta(x, r) \cap C_\delta(y, s)$ is contained in a disc centered at ζ with radius $\lesssim \sqrt{\frac{\Delta + \delta}{d + \delta}}$.

<u>Proof</u> We use the following fact: if $\mu > 0, \epsilon > 0$ then the set

$$\{x \in [-\pi, \pi] : |\cos x - \mu| \leq \epsilon\}$$

is (i) contained in the union of two intervals of length $\lesssim \frac{\epsilon}{\sqrt{|1 - \mu|}}$ and (ii) contained in an interval of length $\lesssim \sqrt{|1 - \mu| + \epsilon}$ centered at 0.

To prove the lemma, we use complex notation and may assume that $x = 0$, $r = 1$, y is on the positive real axis and $s < 1$. Note that then $e = 1$. If $d \leq 4\delta$ then the lemma is trivial, and if $y < \frac{d}{2} - \delta$ then $y + s < 1 - 2\delta$ so that $C_\delta(0, 1) \cap C_\delta(y, s) = \varnothing$. So we can assume that $d \geq 4\delta$ and $y \geq \frac{d}{2} - \delta \geq \frac{d}{4}$.

If $z \in C_\delta(0, 1) \cap C_\delta(y, s)$ then clearly $|z - e^{i\theta}| \leq \delta$ for some $\theta \in [-\pi, \pi]$. It suffices to show that the set of θ which can occur here is contained in two intervals of length $\lesssim \frac{\delta}{\sqrt{(\delta + \Delta)(\delta + d)}}$ and in an interval of length $\lesssim \sqrt{\frac{\Delta + \delta}{d + \delta}}$ centered at 0.

The point $e^{i\theta}$ must belong to $C_{2\delta}(y, s)$, i.e. $\big| |e^{i\theta} - y| - s \big| < 2\delta$ and therefore, since $\big| |e^{i\theta} - y| + s \big| \approx 1$,

$$\big| |e^{i\theta} - y|^2 - s^2 \big| \lesssim \delta.$$

We can express this as

$$\left| \cos \theta - \frac{1 + y^2 - s^2}{2y} \right| \lesssim \frac{\delta}{y} \lesssim \frac{\delta}{d}.$$

Let $\mu = \frac{1+y^2-s^2}{2y}$, $\epsilon = C\frac{\delta}{d}$. Then μ is positive, and

$$|1-\mu| = \frac{|s^2-(1-y)^2|}{2y} \approx \frac{|1-s-y|}{2y} \approx \frac{\Delta}{d}.$$

Apply fact (ii) in the first paragraph. The set of possible θ is therefore contained in an interval of length $\lesssim \sqrt{\frac{\Delta+\delta}{d}}$ centered at 0. This proves (b), since we are assuming $d \geq \delta$. Estimate (a) follows from (b) if $\Delta \leq \delta$. If $\Delta \geq \delta$, then fact (i) in the first paragraph gives the additional property that θ must be contained in the union of two intervals of length $\lesssim \frac{\delta/d}{\sqrt{\Delta/d}} \approx \frac{\delta}{\sqrt{(\delta+\Delta)(\delta+d)}}$.   □

<u>Lemma 3.2</u> (Marstrand's 3-circle lemma) For a suitable numerical constant C_0, assume that $\epsilon, t, \lambda \in (0,1)$ satisfy $C_0\frac{\epsilon}{t} \leq \lambda^2$. Fix three circles $C(x_i, r_i), 1 \leq i \leq 3$. Then for $\delta \leq \epsilon$ the set

$$\overline{\Omega}_{\epsilon t\lambda} \overset{\text{def}}{=} \{(x,r) \in \mathbb{R}^2 \times \mathbb{R} : \Delta((x,r),(x_i,r_i)) < \epsilon \; \forall i,$$
$$d((x,r),(x_i,r_i)) > t \; \forall i, \; C_\delta(x,r) \cap C_\delta(x_i,r_i) \neq \varnothing \; \forall i,$$
$$\text{dist}(C_\delta(x,r) \cap C_\delta(x_i,r_i), C_\delta(x,r) \cap C_\delta(x_j,r_j)) \geq \lambda \; \forall i,j : i \neq j\}$$

is contained in the union of two ellipsoids in $\mathbb{R}^3$ each of diameter $\lesssim \frac{\epsilon}{\lambda^2}$ and volume $\lesssim \frac{\epsilon^3}{\lambda^3}$.

<u>Proof</u> This will be based on the inverse function theorem. We remark that the sketch of proof given in [**61**] is inaccurate.

We will actually work with a slightly different set, namely, with

$$\Omega_{\epsilon t\lambda} = \{(x,r) \in \mathbb{R}^2 \times \mathbb{R} : \Delta((x,r),(x_i,r_i)) < \epsilon \; \forall i, \; d((x,r),(x_i,r_i)) > t \; \forall i,$$
$$|e_i(x,r) - e_j(x,r)| \geq \lambda \; \forall i,j : i \neq j\},$$

where $e_i(x,r) = \text{sgn}(r-r_i)\frac{x_i-x}{|x_i-x|}$. This is sufficient since by Lemma 3.1 (b), $\Omega_{\epsilon t\frac{\lambda}{2}}$ will contain $\overline{\Omega}_{\epsilon t\lambda}$ provided C_0 is sufficiently large.

If $e_1, \dots, e_4$ are unit vectors in $\mathbb{R}^2$ which are contained in an arc of length μ, then the reader will convince herself or himself that

(21) $$|(e_1-e_2) \wedge (e_3-e_4)| \lesssim \mu|e_1-e_2| \, |e_3-e_4|$$

and furthermore if e_1, e_2, e_3 are unit vectors in $\mathbb{R}^2$ then

(22) $$|(e_1-e_2) \wedge (e_1-e_3)| \approx |e_1-e_2| \, |e_2-e_3| \, |e_3-e_1|.$$

Here $\wedge$ is wedge product, $(a,b) \wedge (c,d) = ad - bc$.

Consider the map $G : \mathbb{R}^2 \times \mathbb{R} \to \mathbb{R}^3$ defined by

$$G(x,\rho) = \begin{pmatrix} |x-x_1| - |r-r_1| \\ |x-x_2| - |r-r_2| \\ |x-x_3| - |r-r_3| \end{pmatrix}.$$

Fix $(\xi,\rho) \in \Omega_{\epsilon t\lambda}$. Observe that

(23) $$DG(\xi,\rho) \simeq \begin{pmatrix} e_1(\xi,\rho) & -1 \\ e_2(\xi,\rho) & -1 \\ e_3(\xi,\rho) & -1 \end{pmatrix},$$

where "$\simeq$" means that the two matrices are equal after each row of the matrix on the right hand side is multiplied by an appropriate choice of ± 1.

We can assume that

$$|e_1(\xi,\rho) - e_3(\xi,\rho)| \geq |e_1(\xi,\rho) - e_2(\xi,\rho)| \geq |e_2(\xi,\rho) - e_3(\xi,\rho)|.$$

Let $\mu = |e_1(\xi,\rho) - e_3(\xi,\rho)|$, $\nu = |e_2(\xi,\rho) - e_3(\xi,\rho)|$; then we have $\mu \geq \nu \gtrsim \lambda$ and also $|e_1(\xi,\rho) - e_2(\xi,\rho)| \approx \mu$. It follows by (22) that $|\det DG(\xi,\rho)| \approx \mu^2\nu$. Furthermore, all entries in the cofactor matrix of $DG(\xi,\rho)$ are easily seen to be $\lesssim \mu$. Let $E(\xi,\rho) = \{(x,r) \in \mathbb{R}^2 \times \mathbb{R} : |DG(\xi,\rho)(x-\xi, r-\rho)| < A\epsilon\}$ for an appropriate large constant A which should be chosen before C_0. Then the preceding considerations imply $E(\xi,\rho)$ is an ellipsoid with

$$(24) \qquad\qquad \operatorname{diam}(E(\xi,\rho)) \lesssim \frac{\epsilon}{\mu\nu},$$

$$(25) \qquad\qquad |E(\xi,\rho)| \lesssim \frac{\epsilon^3}{\mu^2\nu}.$$

We claim that if $(x,r) \in E$ then $DG(x,r)DG(\xi,\rho)^{-1} = I + E$, where I is the 3×3 identity matrix and E is a matrix with norm $\leq \frac{1}{100}$, say.

A matrix calculation shows that each entry of $(DG(x,r) - DG(\xi,\rho))DG(\xi,\rho)^{-1}$ has the form $(\det DG(\xi,\rho))^{-1}(e_i(x,r) - e_i(\xi,\rho)) \wedge (e_j(\xi,\rho) - e_k(\xi,\rho))$ for appropriate i, j, k. We will show below that

$$(26) \qquad\qquad |e_i(x,r) - e_i(\xi,\rho)| \lesssim \frac{\epsilon}{t\nu}.$$

If we assume this then the claim may be proved as follows. (26) implies in particular that all the vectors $e_i(x,r)$ and $e_j(\xi,\rho)$ belong to an arc of length $\lesssim \mu$. Accordingly, using (21),

$$|\det DG(\xi,\rho)^{-1}(e_i(x,r) - e_i(\xi,\rho)) \wedge (e_j(\xi,\rho) - e_k(\xi,\rho))|$$
$$\lesssim \mu \, |\det DG(\xi,\rho)^{-1}| \, |e_i(x,r) - e_i(\xi,\rho)| \, |e_j(\xi,\rho) - e_k(\xi,\rho)|$$
$$\lesssim \mu \cdot (\mu^2\nu)^{-1} \cdot \frac{\epsilon}{t\nu} \cdot \mu \leq \frac{\epsilon}{t\nu^2},$$

which is small.

To prove (26) we abbreviate $e_i = e_i(\xi,\rho)$. Fix i and let $e_i^* \in \mathbb{R}^2$ be a unit vector perpendicular to e_i. If we define j and k via $\{i,j,k\} = \{1,2,3\}$, then a little linear algebra shows that $e_i^* = \alpha(e_i - e_j) + \beta(e_i - e_k)$ with $|\alpha| + |\beta| \lesssim \nu^{-1}$. Furthermore, if we let $(v_1, v_2, v_3) = DG(\xi,\rho)(x-\xi, r-\rho)$, then by (23) we have $|(e_i - e_j) \cdot (x-\xi)| = |v_i \pm v_j| \leq 2\epsilon$ and similarly $|(e_i - e_k) \cdot (x-\xi)| \leq 2\epsilon$. We conclude that $|e_i^* \cdot (x-\xi)| \lesssim \frac{\epsilon}{\nu}$, hence $|e_i^* \cdot (x-x_i)| \lesssim \frac{\epsilon}{\nu}$ since $x_i - \xi$ is parallel to e_i. Also $|x - x_i| \geq \frac{t}{2}$ by (24), so

$$\left| e_i^* \cdot \frac{x - x_i}{|x - x_i|} \right| \lesssim \frac{\epsilon}{t\nu}.$$

This implies that for an appropriate choice of $\pm$

$$(27) \qquad\qquad |e_i(x,r) \pm e_i| \lesssim \frac{\epsilon}{t\nu}.$$

Note though that $r - r_i$ and $|x - x_i|$ are nonzero on $E(\xi,\rho)$: this follows from (24), since ϵ is small compared with t so that $|\xi - x_i| \approx t \approx |\rho - r_i|$. So $(x,r) \to e_i(x,r)$ is a continuous function on $E(\xi,\rho)$ and therefore the sign in (27) is independent of (x,r). So (26) holds and the claim is proved.

If A is large enough then the claim implies via the usual proof of the inverse function theorem that G is a diffeomorphism from a subset of $E(\xi,\rho)$ onto a disc of radius 2ϵ, say. In particular, $E(\xi, \rho)$ must contain a point (x, r) with $G(x, r) = 0$. Then $C(x, r)$ is internally tangent to each $C(x_i, r_i)$; note that by (26) and the bound on the diameter of E, we have $(x, r) \in \Omega_{\epsilon \frac{t}{2} \frac{\lambda}{2}}$ and furthermore, by the claim $E(x, r)$ and $E(\xi, \rho)$ are comparable ellipsoids (each is contained in a fixed dilate of the other). Apollonius' theorem implies there are only two possibilities for the circle $C(x, r)$, and we have just seen that (ξ, ρ) must be contained in one of the two $E(x, r)$'s and that they have the proper dimensions. $\qquad\square$

<u>Proposition 3.3</u> For any $p < \frac{8}{3}$ there is an estimate

$$\|M_\delta f\|_q \leq C\delta^{-\frac{1}{2}(\frac{3}{p}-1)}\|f\|_p, \quad q = 2p'.$$

This implies by the proof of Lemma 1.6 that BRK sets have dimension $\geq 2 - \frac{1}{2}(\frac{3}{p} - 1)$ for any $p < \frac{8}{3}$, i.e. dimension $\geq \frac{11}{6}$. Proposition 3.3 was proved (in generalized form) in [**31**]; it is the partial result which corresponds to the bound (i) in the tangency counting problem. The sharp result ((18) with $p = 3$) incorporating the technique from [**17**] is proved in [**61**].

<u>Proof</u> This will be similar to the proof of the $\frac{1}{2} + \alpha$ bound in remark 1.5. The $p = 1$ case is trivial[4] so it suffices to prove the following restricted weak type bound at the endpoint:

$$(28) \qquad \left|\{r \in [\tfrac{1}{2}, 2] : M_\delta \chi_E(r) > \lambda\}\right| \leq C\left(\frac{|E|}{\delta^{\frac{1}{6}}\lambda^{\frac{8}{3}}}\right)^{\frac{6}{5}}.$$

We may assume in proving (28) that the diameter of the set E is less than one. Consequently in defining $M_\delta f$ we may restrict the point x to the disc $D(0, \frac{1}{4})$. Thus it suffices to prove the following.

Assume that $\lambda \in (0, 1]$ and there are M 3δ-separated values $r_j \in [\frac{1}{2}, 2]$ and points $x_j \in D(0, \frac{1}{4})$ such that $|E \cap C_\delta(x_j, r_j)| \geq \lambda |C_\delta(x_j, r_j)|$. Then

$$(29) \qquad M\delta \leq C\left(\frac{|E|}{\delta^{\frac{1}{6}}\lambda^{\frac{8}{3}}}\right)^{\frac{6}{5}}.$$

We can assume that M is large; for M smaller than any fixed constant (29) holds because $M \neq 0$ implies $|E| \gtrsim \lambda \delta$.

To prove (29) we let μ ("multiplicity") be the smallest number with the following property: there are at least $\frac{M}{2}$ values of j such that

$$(30) \qquad \left|E \cap C_\delta(x_j, r_j) \cap \{x : |\{i : x \in C_\delta(x_i, r_i)\}| \leq \mu\}\right| \geq \frac{\lambda}{2}|C_\delta(x_j, r_j)|.$$

The main estimate is

$$(31) \qquad \mu \lesssim M^{\frac{1}{6}}\lambda^{-\frac{5}{3}}.$$

Before proving (31) we introduce some more notation, as follows. For any $t \in [\delta, 1]$ and $\epsilon \in [\delta, 1]$, let

$$a(t, \epsilon) = C_1^{-1}\left(\frac{\delta}{\epsilon}\right)^\alpha \left(\frac{M\delta}{t} + \frac{t}{M\delta}\right)^{-\alpha}.$$

---

[4]The $p = 2$ case was also known prior to [**31**]; it follows from results of Pecher [**41**].

Here α is a sufficiently small positive constant, and C_1 is a positive constant (easily shown to exist) which is large enough that

$$\text{(32)} \qquad\qquad \sum_{\substack{k \geq 0 \\ l \geq 0}} a(2^k \delta, \, 2^l \delta) < 1$$

for all M and δ. Let $\overline{\lambda}(t, \epsilon) = a(t, \epsilon)\frac{\lambda}{2}$, $\overline{\mu}(t, \epsilon) = a(t, \epsilon)\mu$, $\overline{M}(t, \epsilon) = a(t, \epsilon)\frac{M}{2}$. Also, for each $i, j \in \{1, \dots, M\}$ let

$$\text{(33)} \qquad\qquad \Delta_{ij} = \max\big(\delta, \, \big|\,|x_i - x_j| - |r_i - r_j|\,\big|\big)$$

and for each $j \in \{1, \dots, M\}$, $t \in [\delta, 1], \epsilon \in [\delta, 1]$, let

$$S_{t,\epsilon}(x_j, r_j) \overset{\text{def}}{=} \{i : C_\delta(x_j, r_j) \cap C_\delta(x_i, r_i) \neq \varnothing, \, t \leq |r_i - r_j| \leq 2t \text{ and } \epsilon \leq \Delta_{ij} \leq 2\epsilon\},$$
$$A_{t,\epsilon}(x_j, r_j) \overset{\text{def}}{=} \{x \in C_\delta(x_j, r_j) : |\{i \in S_{t,\epsilon}(x_j, r_j) : x \in C_\delta(x_i, r_i)\}| \geq \overline{\mu}(t, \epsilon)\}.$$

<u>Lemma 3.4</u>   There are numbers $t \in [\delta, 1]$ and $\epsilon \in [\delta, 1]$ with the following property:

There are $\geq \overline{M}(t, \epsilon)$ values of j such that $|A_{t\epsilon}(x_j, r_j)| \geq \overline{\lambda}(t, \epsilon) |C_\delta(x_j, r_j)|$.

<u>Proof</u>   This is a routine pigeonhole argument. By the minimality of μ there are at least $\frac{M}{2}$ values of j such that $|\tilde{E}_j| \geq \frac{\lambda}{2}|C_\delta(x_j, r_j)|$ where

$$\tilde{E}_j = E \cap C_\delta(x_j, r_j) \cap \{x : |\{i : x \in C_\delta(x_i, r_i)\}| \geq \mu\}.$$

For any such j and any $x \in \tilde{E}_j$, (32) implies there are $t = 2^k\delta$ and $\epsilon = 2^l\delta$ such that $x \in A_{t\epsilon}(x_j, r_j)$. Consequently, using (32) again, for any such j there are $t = 2^k\delta$ and $\epsilon = 2^l\delta$ such that

$$\text{(34)} \qquad\qquad |A_{t\epsilon}(x_j, r_j)| \geq \overline{\lambda}(t, \epsilon) |C_\delta(x_j, r_j)|.$$

By (32) once more, there must be a choice of t and ϵ such that (34) holds for at least $\overline{M}(t, \epsilon)$ values of j. This finishes the proof.

We fix once and for all a pair (t, ϵ) for which the conclusion of Lemma 3.4 is valid, and will drop the t, ϵ subscripts when convenient, i.e. will denote $\overline{\lambda}(t, \epsilon)$ by $\overline{\lambda}$, etc. We split the proof of (31) into two cases:

(i) $\overline{\lambda} \geq C_2 \sqrt{\frac{\epsilon}{t}}$

(ii) $\overline{\lambda} \leq C_2 \sqrt{\frac{\epsilon}{t}}$

where C_2 is a sufficiently large constant.

In case (i), which is the main case, we let S be the set of M circles in (29), and let $\overline{S}$ be the set of at least $\overline{M}$ circles in Lemma 3.4. Let Q be the set of all quadruples (j, j_1, j_2, j_3) with $C(x_j, r_j) \in \overline{S}$, $C(x_{j_i}, r_{j_i}) \in S$ for $i = 1, 2, 3$ and such that $j_i \in S_{t,\epsilon}(x_j, r_j)$ for each $i \in \{1, 2, 3\}$ and furthermore

$$\text{dist}(C_\delta(x_j, r_j) \cap C_\delta(x_{j_i}, r_{j_i}), C_\delta(x_j, r_j) \cap C_\delta(x_{j_k}, r_{j_k})) \geq C_3^{-1}\overline{\lambda}$$

for all $i, k \in \{1, 2, 3\}$ with $i \neq k$. Here C_3 is a suitable constant which should be chosen before C_2.

We will make two different estimates on the cardinality of Q. On the one hand, the diameter bound in Lemma 3.2 implies that for fixed j_1, j_2, j_3 there are $\lesssim \frac{\epsilon}{\delta}\overline{\lambda}^{-2}$ values of j such that $(j, j_1, j_2, j_3) \in Q$. Also it follows from the definition of Q that there are $\lesssim M \min(M, \frac{t}{\delta})^2$ possible choices for (j_1, j_2, j_3) : there are at most M choices for j_1, and once j_1 is fixed there are $\lesssim \min(M, \frac{t}{\delta})$ possibilities for each of

j_2 and j_3, since $|r_{j_1} - r_{j_i}| \leq |r_{j_1} - r_j| + |r_j - r_{j_i}| \leq 4t$ for $i = 2$ or 3. We conclude that

$$(35) \qquad |Q| \lesssim \frac{\epsilon}{\delta} \overline{\lambda}^{-2} M \min\left(M, \frac{t}{\delta}\right)^2.$$

On the other hand, if we fix j with $C(x_j, r_j) \in \overline{S}$ then (provided C_3 has been chosen large enough) we can find three subsets F_1, F_2, F_3 of $A_{t,\epsilon}(x_j, r_j)$ such that $\text{dist}(F_l, F_m) \geq 2C_3^{-1}\overline{\lambda}$, $l \neq m$, and $|F_l| \gtrsim \delta\overline{\lambda}$ for each l. For fixed l, we let S_l be those indices $i \in S_{t,\epsilon}(x_j, r_j)$ such that $F_l \cap C_\delta(x_i, r_i) \neq \varnothing$. The sets $C_\delta(x_i, r_i)$, $i \in S_l$ must cover F_l at least $\overline{\mu}$ times. So

$$\sum_{i \in S_l} |F_l \cap C_\delta(x_i, r_i)| \gtrsim \overline{\mu}\,\overline{\lambda}\,\delta.$$

For each fixed i we have $|F_l \cap C_\delta(x_i, r_i)| \lesssim \frac{\delta^2}{\sqrt{t\epsilon}}$ by Lemma 3.1(a). Consequently

$$(36) \qquad |S_l| \gtrsim \delta^{-1}\overline{\mu}\,\overline{\lambda}\sqrt{t\epsilon}.$$

It is easy to see using Lemma 3.1(b) that if $i_l \in S_l$ for $l = 1, 2, 3$ then $(j, i_1, i_2, i_3) \in Q$. So

$$|Q| \gtrsim \overline{M}(\delta^{-1}\overline{\mu}\,\overline{\lambda}\sqrt{t\epsilon})^3.$$

If we compare this with (35) we obtain

$$\overline{\mu}^3 \lesssim \frac{\delta^2}{t^{\frac{3}{2}}\epsilon^{\frac{1}{2}}}\overline{\lambda}^{-5}\min\left(M, \frac{t}{\delta}\right)^2 \frac{M}{\overline{M}},$$

or equivalently

$$\mu^3 \lesssim M^{\frac{1}{2}}\lambda^{-5} \cdot \begin{cases} a(t,\epsilon)^{-9}(\frac{\delta}{\epsilon})^{\frac{1}{2}}(\frac{t}{\delta M})^{\frac{1}{2}} & \text{if } M \geq \frac{t}{\delta}, \\ a(t,\epsilon)^{-9}(\frac{\delta}{\epsilon})^{\frac{1}{2}}(\frac{M\delta}{t})^{\frac{3}{2}} & \text{if } M \leq \frac{t}{\delta}. \end{cases}$$

The expression in the brace is bounded by a constant by the definition of $a(t, \epsilon)$, provided $\alpha < \frac{1}{18}$. So we have proved (31) in case (i).

In case (ii), we fix j with $C(x_j, r_j) \in \overline{S}$ and make the trivial estimate

$$|S_{t\epsilon}(x_j, r_j)| \lesssim \min(M, \tfrac{t}{\delta}).$$

It follows that

$$\overline{\mu}\,\overline{\lambda}\,\delta \lesssim \sum_{i \in S_{t\epsilon}(x_j, r_j)} |C_\delta(x_j, r_j) \cap C_\delta(x_i, r_i)| \lesssim \min(M, \frac{t}{\delta})\frac{\delta^2}{\sqrt{t\epsilon}},$$

where we used Lemma 3.1(a). Thus $\overline{\mu} \lesssim \overline{\lambda}^{-1}\sqrt{\frac{t}{\epsilon}}\min(\frac{M\delta}{t}, 1)$. Using the hypothesis (ii) we therefore have

$$\overline{\mu} \lesssim \overline{\lambda}^{-\frac{5}{3}}\left(\frac{t}{\epsilon}\right)^{\frac{1}{6}}\min\left(\frac{M\delta}{t}, 1\right)$$

i.e.

$$\mu \lesssim \lambda^{-\frac{5}{3}} M^{\frac{1}{6}} \cdot \begin{cases} a(t,\epsilon)^{-\frac{8}{3}}(\frac{\delta}{\epsilon})^{\frac{1}{6}}(\frac{t}{\delta M})^{\frac{1}{6}} & \text{if } M \geq \frac{t}{\delta} \\ a(t,\epsilon)^{-\frac{8}{3}}(\frac{\delta}{\epsilon})^{\frac{1}{6}}(\frac{M\delta}{t})^{\frac{5}{6}} & \text{if } M \leq \frac{t}{\delta} \end{cases}$$

The expression in the brace is bounded by a constant provided $\alpha < \frac{1}{16}$, so we have proved (31).

Completion of proof of Proposition 3.3   Let $\tilde{E} = \{i : x \in C_\delta(x_i, r_i)| \leq \mu\}$.
With notation as above we have

$$|E| \geq |\tilde{E}| \geq \mu^{-1} \sum_j |\tilde{E} \cap C_\delta(x_j, r_j)| \gtrsim \mu^{-1} M \lambda \delta \gtrsim \lambda^{\frac{8}{3}} M^{\frac{5}{6}} \delta$$

by (31). Consequently $(M\delta)^{\frac{5}{6}} \lesssim \frac{|E|}{\delta^{\frac{1}{6}} \lambda^{\frac{8}{3}}}$ and the proposition is proved.   □

Further remarks 3.2. We mention some other recent related work. Schlag
[43] found an essentially optimal $L^p \to L^q$ estimate in the context of Bourgain's
theorem. If we define

$$\mathcal{M}_\delta f(x) = \sup_{1 \leq r \leq 2} \frac{1}{|C_\delta(x, r)|} \int_{C_\delta(x,r)} |f|$$

then there is an estimate

$$\forall \epsilon \ \exists C_\epsilon : \|\mathcal{M}_\delta f\|_5 \lesssim C_\epsilon \delta^{-\epsilon} \|f\|_{\frac{5}{2}}$$

and modulo $\delta^{-\epsilon}$ factors all possible $L^p \to L^q$ bounds for $\mathcal{M}_\delta$ follow by interpolation
from this one. Alternate proofs and further related results are in [46], [61] and [44].
On the other hand a number of endpoint questions remain open. The best known
is the restricted weak type $(2, 2)$ version of Bourgain's theorem.

3.3. A more central open question is the so-called local smoothing conjecture
[48], [36] in $2 + 1$ dimensions. See section 4 below. This is a problem "with
cancellation" and likely not susceptible to purely combinatorial methods without
additional input. On the other hand, it would imply (18) with $p = 4$ via the Sobolev
embedding theorem and is therefore close to including some of the results of [17].
This means perhaps that a proof not involving any combinatorics would have to
contain a significant new idea.

3.4. One can give a discrete heuristic for the Kakeya problem analogous to the
one for the BRK sets problem. What follows is an observation of Schlag and the
author.

There is a substantial literature on incidence problems for lines in $\mathbb{R}^3$; these
problems appear to be quite difficult and are largely open. One relevant paper is
Sharir [47], where the following problem is considered:

Let $\{\ell_j\}_{j=1}^N$ be lines in $\mathbb{R}^3$ and define a joint to be a point where three non-
coplanar ℓ_j's intersect. Then how many joints can there be?

If $\mathcal{J}$ is the set of joints then as is discussed in [47] the natural conjecture is
$|\mathcal{J}| \lesssim N^{\frac{3}{2}}$, which would be sharp by taking $\approx \sqrt{N}$ planes parallel to each of three
given planes and considering the lines formed by intersecting two of the planes;
any point where three planes intersect will be a joint. The "easy" bound in this
problem is $|\mathcal{J}| \lesssim N^{\frac{7}{4}}$ which is proved in [14] using a suitable version of (3). The
bound $\forall \epsilon \ \exists C_\epsilon : |\mathcal{J}| \leq C_\epsilon N^{\frac{23}{14} + \epsilon}$ is proved in [47] using similar techniques to [17].

The heuristic is that a bound $|\mathcal{J}| \lesssim N^\alpha$ should imply that (in $\mathbb{R}^3$)

$$\dim(\text{Kakeya}) \geq \frac{\alpha}{\alpha - 1}.$$

Namely, define a μ-fold point in an arrangement of N lines to be a point where at
least μ lines intersect with (say) no more than half of these lines belonging to any
given 2-plane. Then any bound of the form $|\mathcal{J}| \lesssim N^\alpha$ leads to a corresponding
bound $|\mathcal{P}_\mu| \lesssim \left(\frac{N \log \mu}{\mu}\right)^\alpha$ where $\mathcal{P}_\mu$ is the set of μ-fold points. This may be seen
(rigorously) as follows: let $\mathcal{P}_\mu$ be the set of μ- fold points in the arrangement.

Let A be a large constant and take a random sample of the N lines according to the following rule: each line belongs to the sample independently and with probability $\frac{A \log \mu}{\mu}$. Then with high probability the sample has cardinality $\lesssim \frac{N \log \mu}{\mu}$. Furthermore, it is not hard to show that any point of $\mathcal{P}_\mu$ will be a joint for the lines in the sample with probability at least $1 - \mu^{-B}$, where B is large if A is large. It follows that with high probability at least half the points of $\mathcal{P}_\mu$ will be joints for the sample, hence $|\mathcal{P}_\mu| \lesssim (\frac{N \log \mu}{\mu})^\alpha$.

Now the heuristic part of the argument: suppose we have a Kakeya set E with (say, Minkowski) dimension β. Fix δ and take a δ-separated set of directions and a line segment in each direction contained in E; this gives an arrangement of $\approx \delta^{-2}$ lines $\{\ell_j\}$. Let E^δ be the δ-neighborhood of E; thus $|E^\delta| \approx \delta^\beta$, so E^δ is made up of roughly $\delta^{-\beta}$ δ-discs. A typical point in the δ-neighborhood of E should belong to roughly $\delta^{-(3-\beta)}$ δ-neighborhoods of ℓ_j's, since otherwise the "low multiplicity" arguments discussed e.g. in section 2 would show easily that $|E^\delta| \gg \delta^\beta$. Hence if we ignore the distinction between points and δ-discs then we are dealing with an arrangement of δ^{-2} lines with $\delta^{-\beta}$ $\delta^{-(3-\beta)}$-fold points. We conclude that up to logarithmic factors

$$\delta^{-\beta} \lesssim \left(\frac{\delta^{-2}}{\delta^{-(3-\beta)}} \right)^\alpha, \quad \text{i.e.} \quad \beta \geq \frac{\alpha}{\alpha - 1}.$$

Under this heuristic the result of [**47**] would correspond to an improvement over $\frac{5}{2}$ on Kakeya, and the fact that the joints problem is open would seem to indicate that questions 1 and 2 are quite difficult even on a combinatorial level, if in fact the answers are affirmative. In this connection, we note that Schlag [**45**] has proved an analogue of the 3-circle lemma in this context and has used it to give an alternate proof of the result dim(Kakeya) $\geq \frac{7}{3}$ (originally due to Bourgain [**7**]) which corresponds to the result from [**14**] via $\frac{7}{3} = \frac{7/4}{7/4 - 1}$. However, it is not easy to put the argument of [**47**] into the continuum and the author believes that in contrast to the situation considered in [**61**] it may not be possible to do this in a reasonably straightforward way.

A further remark is that special cases of the three dimensional Kakeya problem correspond to results analogous to [**61**] with circles replaced by families of curves satisfying the cinematic curvature condition from [**48**]. For example, the case of sets invariant by rotations around an axis is a problem of this type as is discussed in [**31**].

4. Oscillatory integrals and Kakeya

It seemed appropriate to include a discussion of the basic open problems in harmonic analysis connected with Kakeya, but we will not attempt a complete survey and will not say anything about the proofs of the deeper results. We will just state some well-known open problems and show how they lead to questions 1 and 2.

Let $\hat{f}$ be the Fourier transform and if m is a given function, then let $T_m f$ be the corresponding multiplier operator,

$$\widehat{T_m f} = m \hat{f}.$$

Two longstanding problems in L^p harmonic analysis are the following:

<u>Restriction problem</u> Is there an estimate

(37) $\|\widehat{fd\sigma}\|_p \lesssim \|f\|_{L^p(d\sigma)}$

for all $p > \frac{2n}{n-1}$, where σ is surface measure on the unit sphere $S^{n-1} \subset \mathbb{R}^n$?

<u>Bochner–Riesz problem</u> Let m_δ be a smooth cutoff to a δ-neighborhood of S^{n-1}, i.e.

$$m_\delta(\xi) = \phi(\delta^{-1}(1 - |\xi|)),$$

where $\phi \in C_0^\infty(\mathbb{R})$ is supported in $(-\frac{1}{2}, \frac{1}{2})$. Then is there an estimate

(38) $\forall \epsilon \; \exists C_\epsilon : \|T_{m_\delta} f\|_p \leq \delta^{-\epsilon} \|f\|_p$

when $p \in [\frac{2n}{n+1}, \frac{2n}{n-1}]$?

Both these problems can be formulated in a number of different ways; the formulations we have given are not the original ones but are well-known to be equivalent to them. In fact it would also be equivalent to prove (37) in the weaker form $\|\widehat{fd\sigma}\|_p \lesssim \|f\|_\infty$, $p > \frac{2n}{n-1}$. This is a consequence of the Stein–Nikisin theory as is pointed out in [7], section 6.

A third problem of more recent vintage [48] is

<u>Local smoothing</u> Let u be the solution of the initial value problem for the wave equation in n space dimensions,

$$\Box u = 0, \; u(\,\cdot\,,0) = f, \; \frac{\partial u}{\partial t}(\,\cdot\,,0) = 0$$

Then is there an estimate

(39) $\forall \epsilon > 0 \; \exists C_\epsilon : \|u\|_{L^p(\mathbb{R}^n \times [1,2])} \leq C_\epsilon \|f\|_{p,\epsilon}$

when $p \in [2, \frac{2n}{n-1}]$? Here $\| \cdot \|_{p,\epsilon}$ is the inhomogeneous L^p Sobolev norm with ϵ derivatives.

In all these problems it is well-known that the exponent $\frac{2n}{n-1}$ would be optimal. See [52]. For example, in the last problem this may be seen by considering focussing solutions where f is spread over a δ-neighborhood of the unit sphere and $u(\,\cdot\,,t)$ is mostly concentrated on a δ-disc when $t \in (1, 1 + \delta)$.

When $n = 2$, estimate (37) was proved by Fefferman and Stein and then (38) by Carleson and Sjolin, in the early 1970's (see [52]). Estimate (39) is open even when $n = 2$ however; the known partial results on $L^4(\mathbb{R}^2)$ correspond to loss of $\frac{1}{8}$ derivatives ([36]; an improvement to loss of $\frac{1}{8} - \epsilon$ derivatives appears implicit in [12, p. 60]). In general dimensions, the following implications are known:

$$(39) \Rightarrow (38) \Rightarrow (37) \Rightarrow (2)$$

The first implication is due to Sogge, the second which is deeper is due to Tao [58], and Carbery [13] had shown earlier that the second implication can be reversed in a slightly different context (replace spheres by paraboloids). We refer to [58] for further discussion. Here though we will only be concerned with the last implication which makes the connection with the Kakeya problem. Essentially this is due to Fefferman [23], another relevant reference is [3] and the result as presented here is from [10]. A basic open problem in the area is to what extent the last implication can be reversed. An alternate proof of the two dimensional Carleson–Sjolin result along these lines was given by Cordoba [18]. In three or more dimensions, progress

on this problem was initiated by Bourgain (see [**10**]) who obtained a numerology
between partial results which however does not show that (2) would imply (37).
For a recent improvement in the numerology see [**38**] and [**59**].

A problem of a somewhat different nature is

<u>Montgomery's conjecture</u> Assume $T \leq N^2$. Consider a Dirichlet series

$$D(s) = \sum_{n=1}^{N} a_n n^{is},$$

where $\|\{a_n\}\|_{\ell^\infty} \leq 1$. Let $\mathcal{T}$ be a 1-separated subset of $[0, T]$. Then

$$\forall \epsilon \; \exists C_\epsilon : \sum_{t \in \mathcal{T}} |D(t)|^2 \leq N^\epsilon (N + |\mathcal{T}|) N.$$

An easy consequence (or reformulation) would be that

(40)
$$\forall \epsilon \; \exists C_\epsilon : \int_E |D(t)|^2 dt \leq N^\epsilon (N + |E|) N$$

if $E \subset [0, T]$ with the stated hypotheses on T and $D(s)$. This is an estimate on
the measure of the set of large values of $D(s)$ and would also imply estimates of
L^p norms with $p > 2$. See [**9**] and e.g. [**37**] for these remarks as well as some
discussion of the relationship between (40) and open problems in analytic number
theory. Estimate (40) can perhaps be thought of as an analogue of (37) where the
oscillatory sum operator $\{a_n\} \to D(s)$ replaces the extension operator $f \to \widehat{f d\sigma}$.
Bourgain [**9**] showed that (40) is again related to the Kakeya problem.

In the rest of this article, we will discuss implications of this type, i.e.

oscillatory integral estimates $\Rightarrow$ Kakeya estimates

We first show that (37) implies (2), and will record the corresponding implications between partial results. Let us recall the results that would follow from
(37) using Holder's inequality and interpolation with the trivial bound $\|\widehat{f d\sigma}\|_\infty \leq$
$\|f\|_{L^1(d\sigma)}$, say

(41)
$$\|\widehat{f d\sigma}\|_q \lesssim \|f\|_{L^p(d\sigma)}, \quad p < \frac{2n}{n-1}, \quad q > \frac{n+1}{n-1} p'.$$

This bound for $p \leq 2$ (plus its endpoint version where $q = \frac{n+1}{n-1} p'$) is a well-known
theorem proved by Stein and Tomas in the 1970's and the case $p = q < 2\frac{n+1}{n-1} + \epsilon$
for suitable $\epsilon > 0$ was proved more recently by Bourgain [**7**] using considerations
related to question 2. See [**52**] and [**10**].

<u>Proposition 4.1</u> Assume (41) holds for a given $p \geq 2$ and $q \geq 2$. Then, with
$r = (\frac{q}{2})'$ and $s = (\frac{p}{2})'$, the restricted weak type (r, s) norm of the Kakeya maximal
operator is $\lesssim \delta^{-2(\frac{n}{r} - 1)}$. Consequently the Hausdorff dimension of Kakeya sets is
$\geq 2r - n = \frac{2q}{q-2} - n$. In particular (37) implies (2).

<u>Proof</u> First let $\{T_j\}_{j=1}^N$, $T_j = T_{e_j}^\delta(a_j)$ be any collection of δ-tubes with δ-
separated directions e_j. Let $\tilde{T}_j = \{x \in \mathbb{R}^n : \delta^2 x \in T_j\}$ be the dilation of T_j
by a factor δ^{-2}, and let χ_j and $\tilde{\chi}_j$ be the characteristic functions of T_j and $\tilde{T}_j$
respectively. Let C_j be a spherical cap with radius $\approx \delta$ centered at e_j, e.g. $C_j -$
$\{e \in S^{n-1} : e \cdot e_j \geq 1 - C^{-1}\delta^2\}$ where C is a suitable constant. Take a bump functi

supported in C_j, say $\phi_j \in C_0^\infty(C_j)$ with $\|\phi_j\|_\infty = 1$, $\phi_j \geq 0$ and $\|\phi_j\|_1 \approx \delta^{n-1}$, and let $\psi_j(\xi) = e^{2\pi i \xi \cdot \delta^{-2} a_j} \phi_j(\xi)$. If $x \in \tilde{T}_j$, then the integral

$$\widehat{\psi_j d\sigma}(x) = \int_{S^{n-1}} \psi_j(\xi) e^{-2\pi i \xi \cdot x} d\xi = e^{-2\pi i e_j \cdot (x - \delta^{-2} a_j)} \int_{C_j} \phi_j(\xi) e^{-2\pi i (\xi - e_j) \cdot (x - \delta^{-2} a_j)} d\xi$$

defining $\widehat{\psi_j d\sigma}(x)$ involves no cancellation, so

$$(42) \qquad\qquad\qquad |\widehat{\psi_j d\sigma}| \gtrsim \delta^{n-1} \tilde{\chi}_j.$$

Now consider the function $f = \sum_j \epsilon_j \psi_j$ where the ϵ_j are random ± 1's. Since the supports of the ψ_j are disjoint we have

$$\|f\|_{L^p(S^{n-1})} \lesssim (N\delta^{n-1})^{\frac{1}{p}}$$

and therefore, by the assumption (41),

$$(43) \qquad\qquad\qquad \|\widehat{f d\sigma}\|_q \lesssim (N\delta^{n-1})^{\frac{1}{p}}$$

for any choice of $\pm$. On the other hand, if we let $\mathbb{E}$ denote expectation with respect to the choices of $\pm$, then by Khinchin's inequality and (42)

$$\mathbb{E}(|\widehat{f d\sigma}|^q) \gtrsim \delta^{q(n-1)} \left(\sum_j \tilde{\chi}_j \right)^{\frac{q}{2}}$$

pointwise. If we integrate this inequality and compare with (43) we obtain

$$\delta^{q(n-1)} \left\| \sum_j \tilde{\chi}_j \right\|_{\frac{q}{2}}^{\frac{q}{2}} \lesssim (N\delta^{n-1})^{\frac{q}{p}}.$$

Rescaling by δ^2, then taking $\frac{q}{2}$th roots,

$$\delta^{2(n-1) - \frac{4n}{q}} \left\| \sum_j \chi_j \right\|_{\frac{q}{2}} \lesssim (N\delta^{n-1})^{\frac{2}{p}}.$$

Now let E be a set, $f = \chi_E$ and

$$\Omega = \{e : f_\delta^*(e) \geq \lambda\}.$$

Let $\{e_j\}_{j=1}^N$ be a maximal δ-separated subset of Ω and for each j choose a δ-tube T_j as above with $|E \cap T_j| \geq \lambda |T_j|$. Then

$$N\lambda \delta^{n-1} \leq \sum_j |T_j \cap E| \lesssim |E|^{1-\frac{2}{q}} \left\| \sum_j \chi_j \right\|_{\frac{q}{2}} \lesssim |E|^{1-\frac{2}{q}} (N\delta^{n-1})^{\frac{2}{p}} \delta^{-2(n-1) + \frac{4n}{q}}.$$

Using (6) this implies that

$$|\Omega|^{1-\frac{2}{p}} \lesssim \lambda^{-1} |E|^{1-\frac{2}{q}} \delta^{-2(n-1) + \frac{4n}{q}},$$

i.e. $|\Omega|^{\frac{1}{s}} \lesssim \lambda^{-1} |E|^{\frac{1}{r}} \delta^{-2(\frac{n}{r}-1)}$ which is the bound that was claimed. The dimension statement in the proposition then follows from Lemma 1.6, and the last statement also follows by letting $p \to \frac{2n}{n-1}$ and using well-known formal arguments. $\qquad \square$

Remarks 4.1. The original Fefferman construction was of course a counterexample; essentially he showed

If the disc multiplier were bounded on L^p with $p \neq 2$, then families of tubes with the property in Remark 1.2 could not exist.

The paper [3] applies the argument from [23] to the restriction problem in the above way but the result is again formulated as a counterexample. The formulation as an implication concerning the maximal function is from [10].

4.2. We present another application of the Fefferman construction which shows the following.

<u>Claim</u> For any $n \geq 2, p > 2, K < \infty$, there are solutions of $\Box u = f$ in n space dimensions, with $\|f\|_\infty \leq 1$, $\operatorname{supp} f \subset D(0, 100) \times [0, 1]$, and

$$(44) \qquad \int_2^3 \left\| \frac{\partial u}{\partial t}(\cdot, t) \right\|_{L^p(\mathbb{R}^n)} dt > K.$$

The analogous statement with the x-gradient replacing the t-derivative can be proved in a similar way. The statement can be understood as follows: the energy estimate for the wave equation implies via Duhamel's principle that $\|\nabla u(\cdot, t)\|_2 \lesssim \|f\|_2$ if say $t \in (2, 3)$ and f is supported in $\mathbb{R}^n \times [0, 1]$. The claim says that there can be no such estimate in L^p, $p > 2$, even if one is willing to average in t as in (39) and to restrict to bounded f with compact support. The claim was proved by the author after discussions with S. Klainerman but it is very close to the surface given [23]. The analogous statement for the initial value problem is essentially that (39) fails if the $W^{p\epsilon}$ norm is replaced by the L^p norm on the right hand side; this is a formal consequence of [23] as was probably first observed by Sogge.

The construction below by no means rules out an estimate with loss of ϵ derivatives. In fact the estimate

$$\int_2^3 \left\| \frac{\partial u}{\partial t}(\cdot, t) \right\|_{L^p(\mathbb{R}^n)}^p dt \lesssim \|f\|_{p,\epsilon}^p$$

with $2 < p \leq \frac{2n}{n-1}$ and any $\epsilon > 0$ would follow from (39) via Duhamel.

<u>Proof of the claim</u> If $x \in \mathbb{R}^n$ then we will use the notation $x = (x_1, \bar{x})$, $\bar{x} \in \mathbb{R}^{n-1}$.

For an appropriate constant C and any small enough δ there is a solution of $\Box u = f$ with

$$\|f\|_\infty \leq 1, \quad \operatorname{supp} f \subset \{(x, t) : 0 \leq t \leq 1, \ 0 \leq x_1 \leq 1, \ |\bar{x}| \leq \delta\}$$

and

$$(45) \qquad \left| \frac{\partial u}{\partial t} \right| \geq C^{-1} \quad \text{when } 2 \leq t \leq 3, \ x \in Y^t$$

where Y^t is a subset of $\{x \in \mathbb{R}^n : 2 \leq x_1 \leq 3, \ |\bar{x}| \leq \delta\}$ with measure $\geq C^{-1}\delta^{n-1}$.

This is essentially just the fact that there are high frequency solutions of the wave equation travelling in a single direction tangent to the light cone, which implies we can find f with the indicated support and such that u restricted to $2 \leq t \leq 3$ is also mostly concentrated where $|\bar{x}| \lesssim \delta$. The conclusion then corresponds to conservation of energy.

A rigorous argument can be based on the explicit choice

$$f(x, t) = e^{2\pi i N(x_1 - t)} \phi(x_1) \psi(\delta^{-1}\bar{x}) \chi(t)$$

where N is very large, ϕ, ψ, χ are fixed nonnegative C_0^∞ functions, $\psi(0) = 1$, $\operatorname{supp} \psi \subset D(0, 1)$, $\operatorname{supp} \phi = \operatorname{supp} \chi = [0, 1]$ and ϕ and χ are strictly positive on

$(0, 1)$. Let u be the corresponding solution of the wave equation. Then u is given by the formula

$$u(x, t) = \int e^{2\pi i x \cdot \xi} \frac{\sin(2\pi(t - s)|\xi|)}{2\pi|\xi|} e^{-2\pi i N s} \hat{\phi}(\xi_1 - N)\delta^{n-1}\hat{\psi}(\delta\bar{\xi})\chi(s) \, d\xi \, ds.$$

One can differentiate for t and then evaluate the resulting integral precisely enough to obtain (45) in the region $|x_1 - t| \leq \frac{1}{2}$, $|\bar{x}| \leq C^{-1}\delta$. We omit the details.

If E is a set in space-time then we will use the notation $E^t = \{x \in \mathbb{R}^n : (x, t) \in E\}$. By Remarks 1.2 and 1.3 we can find disjoint δ-tubes $\{T_j\}_{j=1}^M$ in $\mathbb{R}^n$ ($M \approx \delta^{-(n-1)}$) such that the tubes $\tilde{T}_j$ obtained by translating the T_j's by 2 units along their axes are all contained in a set with small measure $a(\delta)$. Let $\Pi_j = T_j \times [0, 1] \subset \mathbb{R}^n \times \mathbb{R}$, and let $\tilde{\Pi}_j = \tilde{T}_j \times [2, 3]$. By the first step of the proof there are functions u_j and f_j, $\Box u_j = f_j$, with f_j supported on Π_j, $\|f_j\|_\infty \leq 1$, and $\left|\frac{\partial u_j}{\partial t}\right| \geq$ const on a subset $Y_j \subset \tilde{\Pi}_j$ which satisfies $|Y_j^t| \approx \delta^{n-1}$ for each $t \in (2, 3)$. Let $Z = \bigcup_j Y_j$; then $|Z^t| \lesssim a(\delta)$ for any $t \in (2, 3)$.

Let $\{\epsilon_j\}$ be random ± 1's. Consider the functions $u = \sum_j \epsilon_j u_j$, $f = \sum_j \epsilon_j f_j$, which satisfy $\Box u = f$. The Π_j's are disjoint, so $\|f\|_\infty \leq 1$ for any choice of ϵ_j's. On the other hand, by Holder's and Khinchin's inequalities, for any fixed $t \in (2, 3)$ we have

$$\mathbb{E}\left(\int_{Z^t} |\frac{\partial u}{\partial t}(x, t)|^p dx\right)^{\frac{2}{p}} \gtrsim a(\delta)^{-(1-\frac{2}{p})} \mathbb{E}\left(\int_{Z^t} |\frac{\partial u}{\partial t}(x, t)|^2 dx\right)$$

$$= a(\delta)^{-(1-\frac{2}{p})} \int_{Z^t} \sum_j |\frac{\partial u_j}{\partial t}(x, t)|^2 dx$$

$$\gtrsim a(\delta)^{-(1-\frac{2}{p})} \sum_j |Y_j^t| \approx a(\delta)^{-(1-\frac{2}{p})},$$

which shows there can be no estimate of the form

$$\left(\int_2^3 \left\|\frac{\partial u}{\partial t}(\cdot, t)\right\|_{L^p(\mathbb{R}^n)}^2 dt\right)^{\frac{1}{2}} \leq C\|f\|_\infty$$

with $p > 2$ when f has support in $D(0, 100) \times [0, 1]$. We then also obtain (44), since an estimate to $L_t^1(L_x^p)$ would imply an estimate to $L_t^2(L_x^q)$ ($\frac{1}{q} = \frac{1}{2}(\frac{1}{2} + \frac{1}{p})$) by interpolation with the energy estimate to $L_t^\infty(L_x^2)$. $\Box$

We now discuss the argument from [9] relating (40) to (2). Bourgain showed there that Montgomery's conjecture if true would imply Kakeya sets have full dimension and a bound like (2) with a different L^p exponent. We reworked the argument a bit for expository reasons and in order to obtain the precise result (40) $\Rightarrow$ (2).

The logic is that (40) implies a Kakeya type statement for arithmetic progressions, which in turn implies (2) for all n. Thus the implication (40) $\Rightarrow$ (2) follows by combining Propositions 4.2 and 4.3 below.

If $\nu \in (0, 1), \beta \in \mathbb{R}$, then we denote

$$P_\nu^\delta(\beta) = \{x \in [0, 1] : |x - (j\nu + \beta)| < \delta \text{ for some } j \in \mathbb{Z}\}$$

i.e. $P_\nu^\delta(\beta)$ is the δ-neighborhood of the arithmetic progression with modulus ν which contains β, intersected with $[0, 1]$.

Proposition 4.2 Assume the conjecture (40). Then for any ϵ there is C_ϵ such that the following holds.

($*$) Fix $\eta \in (0,1)$, $\delta \in (0,\eta)$. Let $E \subset [0,1]$ be such that

$$(46) \qquad \forall \nu \in Y \; \exists \beta \in \mathbb{R} : |P_\nu^\delta(\beta) \cap E| \geq \lambda |P_\nu^\delta(\beta)|$$

where $\lambda \in (0,1]$ satisfies $\lambda \geq C_\epsilon(\frac{\delta^2}{\eta})^{-\epsilon} \cdot \eta$, and where Y is a subset of $(\frac{\eta}{2},\eta)$ with $|Y| \geq \frac{\eta}{100}$. Then

$$|E| \geq C_\epsilon^{-1}\left(\frac{\delta^2}{\eta}\right)^\epsilon \lambda .$$

Proof This will be formally similar to the proof of Proposition 4.1 if one makes the analogy

$$\text{line segment} \longleftrightarrow \text{arithmetic progression}$$
$$\text{spherical cap} \longleftrightarrow \text{interval of integers}$$

Claim 1 Let N and T be as in (40) and let ϵ_0 be a suitable constant. Then, for $\nu \in [\frac{N}{2},N]$ and $\beta \in \mathbb{R}$, the Dirichlet series

$$(47) \qquad d(s) = \sum_{n:|n-[\nu]|\leq \epsilon_0 \frac{N}{\sqrt{T}}} e^{-i\frac{\beta}{[\nu]}(n-[\nu])} n^{is}$$

satisfies

$$|d(s)| \gtrsim \frac{N}{\sqrt{T}}$$

when $s \leq T$ and $\mathrm{dist}(s, 2\pi\nu\mathbb{Z}+\beta) \leq \sqrt{T}$.

Proof This is the "short sum" construction in [9]. Assume at first that $\nu \in \mathbb{Z}$. The Taylor expansion of the logarithm function shows that

$$n^{is} = \nu^{is} e^{is(\frac{n-\nu}{\nu}+\mathcal{O}(\frac{n-\nu}{\nu})^2))},$$

so that

$$e^{-i\frac{\beta}{\nu}(n-\nu)} n^{is} = \nu^{is} e^{i(n-\nu))\frac{s-\beta}{\nu}+is\mathcal{O}((\frac{n-\nu}{\nu})^2)}.$$

Thus the sum (47) involves no cancellation and the bound follows immediately. The general case (i.e. $\nu \notin \mathbb{Z}$) follows by replacing ν by $[\nu]$ and noting that this does not significantly affect the hypothesis on s, since if $\mathrm{dist}(s, 2\pi\nu\mathbb{Z}+\beta) \leq \sqrt{T}$ then $\mathrm{dist}(s, 2\pi[\nu]\mathbb{Z}+\beta) \leq \sqrt{T} + C\frac{T}{N} \lesssim \sqrt{T}$.

We therefore define $\tilde{P}_\nu(\beta) = \{x \in [0,T] : \mathrm{dist}(s, 2\pi\nu\mathbb{Z}+\beta) \leq \sqrt{T}\}$. We also fix a number $\epsilon > 0$ and let C_ϵ be a suitable constant.

Claim 2 Assume (40) and let E be a subset of $[0,T]$ with the following property: there is a set $Y \subset [\frac{N}{2},N]$ with $|Y| \geq \frac{N}{100}$, such that for any $\nu \in Y$ there is $\beta = \beta(\nu) \in \mathbb{R}$ such that $|E \cap \tilde{P}_\nu(\beta)| \geq \lambda |\tilde{P}_\nu(\beta)|$. Then

$$(48) \qquad |E| \geq C_\epsilon^{-1} N^{-\epsilon} T \lambda$$

provided $\lambda \geq C_\epsilon N^\epsilon \frac{N}{T}$.

<u>Proof</u> Let ϵ_0 be as in claim 1, choose a maximal $2\epsilon_0 \frac{N}{\sqrt{T}} + 1$-separated subset $\{\nu_j\}_{j=1}^M \subset Y$, denote $\tilde{P}_j = \tilde{P}_{\nu_j}(\beta_j)$ and let χ_j be the characteristic function of $\tilde{P}_j$. Construct Dirichlet series

$$d_j(s) = \sum_{n: |n - [\nu_j]| \le \epsilon_0 \frac{N}{\sqrt{T}}} a_n n^{is}$$

via claim 1 so that $|d_j(s)|^2 \gtrsim \frac{N^2}{T}\chi_j$. Let $D(s) = \sum_j \epsilon_j d_j(s)$ where the ϵ_j are random ± 1's. By Khinchin's inequality

$$(49) \qquad\qquad \mathbb{E}(|D(s)|^2) \gtrsim \frac{N^2}{T} \sum_{j=1}^M \chi_j$$

pointwise. On the other hand the coefficient intervals for the d_j are disjoint so for any choice of ± 1, $D(s)$ will be a Dirichlet series with coefficients bounded by 1. Integrating (49) over E and using (40), we obtain

$$\frac{N^2}{T} \sum_{j=1}^M |E \cap \tilde{P}_j| \lesssim \mathbb{E}\left(\int_E |D(s)|^2 \right) \lesssim N^\epsilon (N + |E|)\, N.$$

We have $M \approx \sqrt{T}$, and for each j we have $|E \cap \tilde{P}_j| \ge \lambda \frac{T^{\frac{3}{2}}}{N}$. So we obtain $T\lambda \lesssim N^\epsilon(N + |E|)$. Under the stated hypothesis on λ this implies (48).

Proposition 4.2 follows from claim 2 by rescaling: set $T = \delta^{-2}$ and $N = \eta\delta^{-2}$, and make the change of variables $x \to Tx$, $\nu \to T\nu$. $\qquad\qquad\square$

Proposition 4.3 If $(*)$ holds then (2) holds in all dimensions n.

<u>Proof</u> We first observe that $(*)$ implies a generalization of itself via a well-known formal argument (one of the arguments in the Stein–Nikisin theory, see [**50**, p. 146]). Namely, drop the hypothesis $|Y| \ge \frac{\eta}{100}$. Then, with the other hypotheses unchanged,

$$(50) \qquad\qquad |E| \gtrsim \lambda \frac{|Y|}{\eta} \left(\frac{\delta^2}{\eta} \right)^\epsilon.$$

To prove (50), let ρY be the dilation of Y by ρ. One can find numbers $\{\rho_j\}_{j=1}^M \subset (\frac{1}{2}, 2)$, where $M \approx \frac{\eta}{|Y|}$, so that $\tilde{Y} \overset{\text{def}}{=} \bigcup_j \rho_j Y$ satisfies $|\tilde{Y}| \ge \frac{\eta}{10}$. Let $\tilde{E} = \bigcup_j \rho_j E$. Then $\tilde{E}$ satisfies (46) when $\nu \in \tilde{Y}$ so $|\tilde{E}| \gtrsim \lambda(\frac{\delta^2}{\eta})^\epsilon$, hence $|E| \gtrsim \lambda M^{-1}(\frac{\delta^2}{\eta})^\epsilon$, which is (50).

Now we consider the Kakeya problem, and will give without detailed proof a few reductions made in [**7**, p. 152].

A. In order to prove (2) it suffices to prove the following inequality: let E be a set in $\mathbb{R}^n$, let Ω be a subset of S^{n-1} with $|\Omega| \ge \frac{1}{2}$, and assume that for any $e \in \Omega$ there is a tube $T_e^\delta(a)$ such that $|T_e^\delta(a) \cap E| \ge \lambda |T_e^\delta(a)|$. Then

$$(51) \qquad\qquad \forall \epsilon > 0\ \exists C_\epsilon : \ |E| \ge C_\epsilon^{-1} \delta^\epsilon \lambda^n.$$

To make this reduction one first observes that (2) is equivalent to the corresponding restricted weak type statement,

$$(52) \qquad\qquad \left| \{ e \in S^{n-1} : f_\delta^*(e) \ge \lambda \} \right| \lesssim \delta^{-\epsilon} \frac{|E|}{\lambda^n},$$

where $f = \chi_E$, and then uses the above argument from [**50**] to reduce (52) to the case where the left hand side is $\geq \frac{1}{2}$. Furthermore, if $|E \cap T_e^\delta(a)| \geq \lambda |T_e^\delta(a)|$ even for one choice of e and a then clearly $|E| \gtrsim \lambda \delta^{n-1}$. It follows that in proving (51) we can assume $\lambda \geq \delta$.

B. We define $\mathcal{Q}$ to be the unit cube $[0,1) \times \cdots \times [0,1)$. Let N be an integer to be fixed below, such that $\frac{1}{N} < \delta$. If $\nu \in \mathbb{Z}^n$, then we define Q_ν to be the cube $[\frac{\nu_1}{N}, \frac{\nu_1+1}{N}) \times \cdots \times [\frac{\nu_n}{N}, \frac{\nu_n+1}{N})$. When we refer below to a $\frac{1}{N}$-cube we always mean a cube which is of the form Q_ν for some $\nu \in \mathbb{Z}^n$. In proving (51) we can assume that E is contained in $\mathcal{Q}$; this follows easily since the tubes $T_e^\delta(a)$ have diameter $\lesssim 1$. Furthermore we can assume that E is a union of $\frac{1}{N}$-cubes; see [**7**].

C. It is easy to see that $f_\delta^*(e') \leq C f_\delta^*(e)$ if $|e - e'| \leq \delta$, since any tube $T_{e'}^\delta(b)$ can be covered by a bounded number of tubes of the form $T_e^\delta(a)$. Accordingly if Ω is as in A., C_1 is a constant, and if $\text{dist}(e, \Omega) \leq C_1 \delta$ then there is a such that $|T_e^\delta(a) \cap E| \geq C^{-1}\lambda |T_e^\delta(a)|$ where C depends on C_1.

In proving (51) we may assume that

$$|\Omega \cap \{e \in S^{n-1} : e_1 \geq \tfrac{1}{2}\}|$$

is bounded below by a constant depending on n only, since we can always achieve this by an appropriate choice of coordinates. In addition, as indicated above we may assume $\lambda \geq \delta$, and we may certainly assume that ϵ is small. Fix integers N and B satisfying the following relations:

$$(53) \qquad B^{-1}N^{2n\epsilon} \approx \lambda \text{ and } \frac{B}{N} \approx \delta.$$

Then $N \approx (\delta\lambda)^{\frac{-1}{1-2n\epsilon}}$, so that

$$(54) \qquad N\delta \text{ is large, } N \leq \delta^{-3}, B \text{ is large, and } BN^{-n} \leq B^{-1}.$$

Define a map $\Phi : \mathbb{R}^n \to \mathbb{R}$ via

$$\Phi(x) = \frac{[Nx_1]}{N} + \frac{[Nx_2]}{N^2} + \cdots + \frac{[Nx_{n-1}]}{N^{n-1}} + \frac{Nx_n}{N^n}.$$

Then Φ maps $\mathcal{Q}$ into $[0,1)$. We make a few additional remarks about the definition:

(i) Note the distinguished role played by the last coordinate.

(ii) Φ maps $\frac{1}{N}$-cubes on intervals of length N^{-n}, hence if E is a union of $\frac{1}{N}$-cubes then $|\Phi(E)| = |E|$.

(iii) Suppose that $x \in \mathbb{R}^n$. Then x belongs to a unique $\frac{1}{N}$-cube Q_ν. Define $\tau(x)$ ("tower over x") via

$$\tau(x) = \bigcup(Q_\mu : \mu_j = \nu_j \text{ when } j < n \text{ and } |\mu_n - \nu_n| \leq B).$$

Then, for any x, Φ maps $\tau(x)$ on an interval of length $\frac{2B+1}{N^n}$.

(iv) Suppose that $w = (\frac{k_1}{N}, \ldots, \frac{k_n}{N})$ where the $\{k_j\}$ are integers. Set $\nu(w) = \sum_j \frac{k_j}{N^j}$. Then Φ maps any arithmetic progression $\{x + jw\}_{j\in\mathbb{Z}}$ to an arithmetic progression in $\mathbb{R}$ with modulus $\nu(w)$.

A lattice vector will be by definition a vector in $\mathbb{R}^n$ of the form

$$w = \left(\frac{k_1}{N}, \ldots, \frac{k_n}{N}\right),$$

where the $\{k_j\}$ are integers with $k_1 \in (\frac{N}{2B}, \frac{N}{B})$ and $\sqrt{\sum_j k_j^2} \leq 2k_1$. Thus any lattice vector w satisfies $|w| \approx \frac{1}{B}$. We note that if $e \in S^{n-1}$ satisfies $e_1 \geq \frac{1}{2}$ then $|e - \frac{w}{|w|}| \lesssim \delta$ for approximately $\frac{N}{B}$ lattice vectors w, namely all the lattice vectors $w = \frac{k}{N}$ which correspond to integer vectors k such that $|k - te| \lesssim 1$ for some t with $t \approx \frac{N}{B}$. Accordingly, for an appropriate constant A there are $\gtrsim (\frac{N}{B})^n$ lattice vectors w such that $\mathrm{dist}(\frac{w}{|w|}, \Omega) \leq A\delta$. We denote this set of lattice vectors by Λ.

If $w \in \Lambda$, then we will abuse our notation slightly and denote the tube $T^\delta_{\frac{w}{|w|}}(a)$ by $T^\delta_w(a)$. By C. above, for each $w \in \Lambda$ we can choose $a \in \mathbb{R}^n$ so that $|T^\delta_w(a) \cap E| \gtrsim \lambda |T^\delta_w(a)|$. It then follows by an averaging argument[5] that there is $a' \in \mathbb{R}^n$ such that

$$(55) \qquad \left| E \cap \left(\bigcup_{j=1}^B \tau(a' + jw) \right) \right| \gtrsim \lambda \left| \bigcup_{j=1}^B \tau(a' + jw) \right|$$

Now set $\rho = \frac{B}{4N^n}$. By (iv) above, the image of the progression a', $a' + w$, $\ldots$, $a' + Bw$ under Φ is an arithmetic progression β, $\beta + \nu(w)$, $\ldots$, $\beta + B\nu(w)$. By (iii), $\Phi(\bigcup_{j=1}^B \tau(a' + jw))$ is a union of intervals containing the points of this progression, with the length of each interval being less than ρ and comparable to ρ. Since E and $\bigcup_{j=1}^B \tau(a' + jw)$ are unions of $\frac{1}{N}$-cubes, (55) and (ii) then imply that $|\Phi(E) \cap P^\rho_\nu(\beta)| \gtrsim \lambda |P^\rho_\nu(\beta)|$. We conclude:

If $\nu = \nu(w)$ for some $w \in \Lambda$, then there is β such that

$$(56) \qquad |P^\rho_\nu(\beta) \cap E| \gtrsim \lambda |P^\rho_\nu(\beta)|$$

Let $Y = \{\nu \in \mathbb{R} : |\nu - \nu(w)| \leq N^{-n} \text{ for some } w \in \Lambda\}$. It follows easily that (56) continues to hold (for suitable β) for any $\nu \in Y$. Note that $Y \subset (\frac{1}{2B}, \frac{2}{B})$ (because of the requirement $\frac{N}{2B} \leq k_1 \leq \frac{N}{B}$) and also $|Y| \gtrsim B^{-n}$, since the set $\{\nu(w) : w \in \Lambda\}$ is N^{-n}-separated and has cardinality $\gtrsim (\frac{N}{B})^n$.

Now λ is large compared with $B^{-1} \cdot (B(BN^{-n})^2)^{-\epsilon}$ by (53), (54), so we can apply (50) with $\eta = B^{-1}$, $\delta = BN^{-n}$, and with $\frac{|Y|}{\eta} \gtrsim B^{-(n-1)}$. We conclude that $|\Phi(E)| \gtrsim \lambda B^{-(n-1)}(B^3 N^{-2n})^\epsilon$. Again using (53) and (54), we obtain $|\Phi(E)| \gtrsim \lambda^n N^{-2n^2\epsilon} \geq \lambda^n \delta^{6n^2\epsilon}$. But E is a union of $\frac{1}{N}$-cubes so $|E| = |\Phi(E)|$, and since ϵ is arbitrary this proves (51). $\square$

References

[1] D. Alvarez, Berkeley thesis, 1997, and to appear.

[2] J. Barrionuevo, A note on the Kakeya maximal operator, Math. Research Letters 3 (1996), 61–65.

[3] W. Beckner, A. Carbery, S. Semmes, F. Soria, A note on restriction of the Fourier transform to spheres, Bull. London Math. Soc. 21 (1989), 394–398.

[4] A.S. Besicovitch, R. Radó, A plane set of measure zero containing circumferences of every radius, J. London Math. Soc. 43 (1968), 717–719.

---

[5]Namely: let m be the measure of the set $\bigcup_{j=1}^B \tau(a' + jw)$; m is clearly independent of a', and furthermore if $x \in \mathbb{R}^n$ is given then the measure of the set $\sigma_x = \{a' : x \in \bigcup_{j=1}^B \tau(a' + jw)\}$ is also comparable to m. If $x \in T^\delta_w(a)$ then, since $\frac{B}{N} \lesssim \delta$ and $|w| \lesssim \frac{1}{B}$, the set σ_x will be contained in $\tilde{T}^\delta_w(a)$, the dilation of $T^\delta_w(a)$ by a suitable fixed constant. It follows that

$$\int_{\tilde{T}^\delta_w(a)} |E \cap (\bigcup_{j=1}^B \tau(a' + jw))| \, da' \geq \int_{T^\delta_w(a) \cap E} |\sigma_x| \, dx \geq \lambda m |T^\delta_w(a)| \approx \lambda m |\tilde{T}^\delta_w(a)|,$$

so (55) holds for suitable $a' \in \tilde{T}^\delta_w(a)$.

[5] B. Bollobas, *Graph Theory: an introductory course*, Graduate Texts in Mathematics vol. 63, Springer-Verlag, 1979.

[6] J. Bourgain, Averages in the plane over convex curves and maximal operators, J. Analyse Math. 47 (1986), 69–85.

[7] J. Bourgain, Besicovitch type maximal operators and applications to Fourier analysis, Geometric and Functional Analysis 1 (1991), 147–187.

[8] J. Bourgain, L^p estimates for oscillatory integrals in several variables, Geometric and Functional Analysis 1 (1991), 321–374.

[9] J. Bourgain, On the distribution of Dirichlet sums, J. Anal. Math. 60 (1993), 21–32.

[10] J. Bourgain, Some new estimates for oscillatory integrals, in *Essays on Fourier analysis in honor of Elias M. Stein*, ed. C. Fefferman, R. Fefferman, S. Wainger, Princeton University Press, 1994.

[11] J. Bourgain, Hausdorff dimension and distance sets, Israel J. Math 87 (1994), 193–201.

[12] J. Bourgain, Estimates for cone multipliers, Operator Theory: Advances and Applications, 77 (1995), 41–60.

[13] A. Carbery, Restriction implies Bochner–Riesz for paraboloids, Proc. Cambridge Phil. Soc. 111 (1992), 525–529.

[14] B. Chazelle, H. Edelsbrunner, L. J. Guibas, R. Pollack, R. Seidel, M. Sharir, J. Snoeyink, Counting and cutting cycles of lines and rods in space, Comput. Geom. Theory Appls. 1 (1992), 305- 323.

[15] M. Christ, Estimates for the k- plane transform, Indiana Univ. Math. J. 33 (1984), 891–910.

[16] M. Christ, J. Duoandikoetxea, J. L. Rubio de Francia, Maximal operators related to the Radon transform and the Calderón–Zygmund method of rotations, Duke Math. J. 53 (1986),189–209.

[17] K. L. Clarkson, H. Edelsbrunner, L. J. Guibas, M. Sharir, E. Welzl, Combinatorial complexity bounds for arrangements of curves or spheres, Discrete Comput. Geom. 5 (1990), 99–160.

[18] A. Cordoba, The Kakeya maximal function and spherical summation multipliers, Amer. J. Math. 99 (1977), 1–22.

[19] R. O. Davies, Some remarks on the Kakeya problem, Proc. Cambridge Phil. Soc. 69 (1971), 417–421.

[20] S. Drury, L^p estimates for the x-ray transform, Ill. J. Math. 27 (1983), 125–129.

[21] K. J. Falconer, *The geometry of fractal sets*, Cambridge University Press, 1985.

[22] K. J. Falconer, On the Hausdorff dimension of distance sets, Mathematika 32 (1985), 206–212.

[23] C. Fefferman, The multiplier problem for the ball, Ann. Math. 94 (1971), 330–336.

[24] R. L. Graham, B. L. Rothschild, J. H. Spencer, *Ramsey Theory*, 2nd edition, Wiley- Interscience, 1990.

[25] M. de Guzman, *Real variable methods in Fourier Analysis*, North-Holland, 1981.

[26] N. Katz, A counterexample for maximal operators over a Cantor set of directions, Math. Research Letters 3 (1996), 527–536.

[27] N. Katz, Remarks on maximal operators over arbitrary sets of directions, to appear.

[28] N. Katz, to appear.

[29] U. Keich, to appear

[30] J. R. Kinney, A thin set of circles, Amer. Math. Monthly 75 (1968), 1077–1081.

[31] L. Kolasa, T. Wolff, On some variants of the Kakeya problem, Pac. J. Math., to appear.

[32] L. Kuipers, H. Niederreiter, *Uniform Distribution of Sequences*, Wiley-Interscience, 1974.

[33] J. M. Marstrand, Packing planes in $\mathbb{R}^3$, Mathematika 26 (1979), 180–183.

[34] J. M. Marstrand, Packing circles in the plane, Proc. London Math. Soc. 55 (1987), 37–58.

[35] W. Minicozzi, C. Sogge, Negative results for Nikodym maximal functions and related oscillatory integrals in curved space, to appear.

[36] G. Mockenhoupt, A. Seeger, C. Sogge, Wave front sets and Bourgain's circular maximal theorem, Ann. Math. 134 (1992), 207–218.

[37] H. L. Montgomery, *Ten lectures on the interface between analytic number theory and harmonic analysis*, CBMS Regional Conference Series in Mathematics, vol. 84, American Mathematical Society, 1994.

[38] A. Moyua, A. Vargas, L. Vega, Schrodinger maximal functions and restriction properties of the Fourier transform, International Math. Res. Notices no. 16 (1996), 793–815.

[39] D. M. Oberlin, E. M. Stein, Mapping properties of the Radon transform, Indiana Univ. Math. J. 31 (1982), 641–650.

[40] J. Pach, P. Agarwal, *Combinatorial Geometry*, Wiley-Interscience, 1995.

[41] H. Pecher, Nonlinear small data scattering for the wave and Klein–Gordon equation, Math. Z. 185 (1984), 261–270.

[42] E. Sawyer, Families of plane curves having translates in a set of measure zero, Mathematika 34 (1987), 69–76.

[43] W. Schlag, A generalization of Bourgain's circular maximal theorem, J. Amer. Math. Soc. 10 (1997), 103–122.

[44] W. Schlag, A geometric proof of the circular maximal theorem, Duke Math. J., to appear.

[45] W. Schlag, A geometric inequality with applications to the Kakeya problem in three dimensions, Geometric and Functional Analysis, to appear.

[46] W. Schlag, C. Sogge, Local smoothing estimates related to the circular maximal theorem, Math. Research Letters 4 (1997), 1–15.

[47] M. Sharir, On joints in arrangements of lines in space, J. Comb. Theory A 67 (1994), 89–99.

[48] C. Sogge, Propagation of singularities and maximal functions in the plane, Inv. Math. 104 (1991), 349–376.

[49] C. Sogge, Concerning Nikodym-type sets in 3-dimensional curved space, to appear.

[50] E. M. Stein, On limits of sequences of operators, Ann. Math. 74 (1961), 140–170.

[51] E. M. Stein, Maximal functions: spherical means, Proc. Nat. Acad. Sci. USA 73 (1976), 2174–2175.

[52] E. M. Stein, *Harmonic Analysis*, Princeton University Press, 1993.

[53] E. M. Stein, G. L. Weiss, *Introduction to Fourier analysis on Euclidean spaces*, Princeton University Press, 1971.

[54] E. M. Stein, N. J. Weiss, On the convergence of Poisson integrals, Trans. Amer. Math. Soc. 140 (1969), 34–54.

[55] L. Szekely, Crossing numbers and hard Erdős problems in discrete geometry, Comb. Prob. Comput. 6 (1997), 353–358.

[56] E. Szemeredi, W. T. Trotter Jr., Extremal problems in discrete geometry, Combinatorica 3 (1983), 381–392.

[57] M. Talagrand, Sur la measure de la projection d'un compact et certaines familles de cercles, Bull. Sci. Math. (2) 104 (1980), 225–231.

[58] T. Tao, The Bochner–Riesz conjecture implies the restriction conjecture, to appear.

[59] T. Tao, A. Vargas, L. Vega, A bilinear approach to the restriction and Kakeya conjectures, to appear.

[60] T. Wolff, An improved bound for Kakeya type maximal functions, Revista Math. Iberoamericana 11 (1995), 651–674.

[61] T. Wolff, A Kakeya type problem for circles, Amer. J. Math. 119 (1997), 985–1026.

[62] T. Wolff, A mixed norm estimate for the x-ray transform, to appear in Revista Math. Iberoamericana.

DEPARTMENT OF MATHEMATICS, 253-37 CALTECH, PASADENA, CA 91125, USA
E-mail address: wolff@cco.caltech.edu